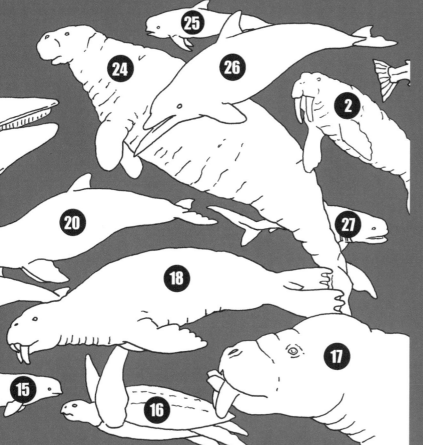

SHARKTOOTH HILL COMMUNITY

1. *Oncorhynchus rastrosus*—saber-toothed salmon
2. *Valenictus imperialensis*—odobenine walrus
3. *Zarhinocetus errabundus*—long-snouted dolphin
4. *Peripolocetus vexillifer*—stem right whale
5. *Carcharocles megalodon*—giant shark
6. *Mola mola*—ocean sunfish
7. *Odontaspis* sp.—sand tiger shark
8. *Imagotaria downsi*—"imagotarine" walrus
9. *Pelagiarctos thomasi*—"imagotarine" walrus
10. *Aulophyseter morricei*—early sperm whale
11. *Allodesmus kernensis*—desmatophocid
12. *Semicossyphus pulcher*—sheepshead wrasse
13. *Neoparadoxia cecilialina*—desmostylian
14. *Desmostylus hesperus*—desmostylian
15. *Denebola brachycephala*—early beluga whale
16. *Psephophorus californiensis*—giant leatherback turtle
17. *Gomphotaria pugnax*—dusignathine walrus
18. *Dusignathus seftoni*—dusignathine walrus
19. *Atocetus nasalis*—kentriodontid
20. *Parapontoporia sternbergi*—extinct "river" dolphin
21. *Balaenoptera bertae*—extinct minke whale relative
22. *Eschrichtius* sp.—gray whale
23. *Makaira nigricans*—extinct marlin
24. *Hydrodamalis cuestae*—giant sea cow
25. *Protoglobicephala mexicana*—early pilot whale
26. *Semirostrum ceruttii*—skim-feeding porpoise
27. *Megachasma* sp.—megamouth shark

The Rise of

Marine Mammals

Annalisa Berta

Graphics Editor James L. Sumich

Illustrations by Carl Buell, Robert Boessenecker,
William Stout, and Ray Troll

The Rise of
Marine Mammals

50 Million Years of Evolution

Johns Hopkins University Press   Baltimore

© 2017 Johns Hopkins University Press
All rights reserved. Published 2017
Printed in China on acid-free paper
9 8 7 6 5 4 3 2 1

Johns Hopkins University Press
2715 North Charles Street
Baltimore, Maryland 21218-4363
www.press.jhu.edu

Library of Congress Cataloging-in-Publication Data

Names: Berta, Annalisa, author. | Sumich, James L., graphics
    editor.
Title: The rise of marine mammals : 50 million years of evolu-
    tion / Annalisa Berta ; graphics editor James L. Sumich.
Description: Baltimore : Johns Hopkins University Press,
    [2017] | Includes bibliographical references and index.
Identifiers: LCCN 2016049243| ISBN 9781421423258 (hard-
    cover : alk. paper) | ISBN 9781421423265 (electronic)
Subjects: LCSH: Marine animals—Evolution. | Mammals—
    Evolution.
Classification: LCC QL121 .B47 2017 | DDC 591.77—dc23
    LC record available at https://lccn.loc.gov/2016049243

A catalog record for this book is available from the British
Library.

Endsheet illustration: The Sharktooth Hill Community.
Illustration by Ray Troll.

Title page illustration: Marine mammals of San Diego's Pliocene
bay. *Clockwise from top:* right whale, balaenopterid attacking
a bait ball of schooling fish, sperm whale, gray whale, beluga,
and Chula Vista walrus. Illustrated by and provided courtesy
of William Stout.

*Special discounts are available for bulk purchases of this book. For
more information, please contact Special Sales at 410-516-6936 or
specialsales@press.jhu.edu.*

Johns Hopkins University Press uses environmentally friendly
book materials, including recycled text paper that is composed
of at least 30 percent post-consumer waste, whenever
possible.

What's past is prologue.

—Shakespeare, *The Tempest*

Contents

Preface

My interest in paleontology developed more than 50 years ago when I participated in summer paleontology excavations at the Clarno Mammal Quarry at Camp Hancock (Oregon Museum of Science and Industry) in central Oregon. The thrill of discovering the bones and teeth of fossil mammals—horses, brontotheres, tapirs, and carnivores—more than 40 million years old fueled my lifelong passion and interest in deciphering the history of life. During this time I had the privilege of working with graduate students Ron Wolff and Bruce Hansen, then at the University of California, Berkeley. After I completed my undergraduate degree with courses in biology and geology, they encouraged me to enroll in graduate school in one of the only places in the United States where one could obtain a PhD in paleontology, the University of California, Berkeley. As a graduate student I studied carnivore evolution and systematics, with a focus on collections-based research. On completing my PhD and after a two-year postdoc at the Florida Museum of Natural History, I took an academic position in the Department of Biology at San Diego State University, where I am fortunate to have been employed since 1982.

A fortuitous meeting with vertebrate paleontologist Clayton Ray at the Smithsonian several years later led to our description of a skeleton of the fossil pinnipedimorph *Enaliarctos mealsi*. Later, with Clayton's encouragement and funding from the National Science Foundation, I began working on the vast collection of well-preserved fossil pinnipeds housed in the Smithsonian's Department of Vertebrate Paleontology, which was amassed by the prodigious amateur fossil collector Douglas Emlong. I also began studying another large collection of fossil marine mammals at the San Diego Natural History Museum that were acquired and curated by vertebrate paleontologist Tom Deméré, thus beginning a research collaboration that has continued for more than 30 years. In the mid-1990s I began collaborating with functional anatomist Ted Cranford, and together, Ted, Tom, and I mentored more than 25 SDSU graduate students' MS theses on the anatomy, evolution, and systematics of fossil as well as living marine mammals, especially pinnipeds and whales.

This book celebrates fossil marine mammals—the extraordinary discoveries and the scientists whose research on these animals has enriched our understanding of their origins, evolution, and diversification. I hope that this book inspires the next generation of marine mammal paleontologists to continue this journey of discovery, collection, and study of these magnificent mammals of the sea. Although much progress in marine mammal paleontology has been made, there is still much left to learn using the latest techniques of study, including 3-D imaging and molecular, finite element, and morphometric analyses, together with classic anatomical study based on detailed drawings and description. Perhaps, most importantly, I want to encourage alliances with colleagues and students in related disciplines such as developmental biology, molecular biology, genetics, geology, and ecology. I have come to appreciate that an integrative approach to paleontology offers valuable opportunities to address research questions from new and diverse perspectives. Finally, Shakespeare's words "What's past is prologue," from *The Tempest*, seem particularly appropriate given the framework of this book, since it suggests that the evolutionary history of marine mammals in many ways sets the stage for what happens to them now and in the future.

Acknowledgments

My long-time collaborator Jim Sumich is acknowledged for providing the excellent line drawings as well as modifying and organizing the artwork. Jim and I first collaborated on a marine mammal textbook nearly 20 years ago, now in its third edition, and, as always, his advice and ability to accurately and creatively render information visually is most appreciated.

I thank former and recent graduate students—Peter Adam, Will Ary, Celia Barroso, Bridget Borce, Morgan Churchill, Lisa Cooper, Liliana Fajardo-Mellor, Giacomo Franco, Reagan Furbish, Carrie Fyler, Anders Galatius, Francis Johnson, Cassie Johnston, Mandy Keogh, Sarah Kienle, Agnese Lanzetti, Jessica Martin, Michael McGowen, Megan McKenna, Sharon Messenger, Rachel Racicot, Amanda Rychel, Alex Sanchez, Meghan Smallcomb, Breda Walsh, Josh Yonas, Samantha Young, and Nick Zellmer—and former postdoc Eric Ekdale, who have worked in my lab at San Diego State University, for their motivation, effort, and enthusiasm for research.

For providing images and the use of artwork I sincerely thank Eli Amson, Tatsuro Ando, Brian Beatty, Giovanni Bianucci, Michelangelo Bisconti, Robert Boessenecker, Carl Buell, Brian Choo, Lisa Cooper, Tom Deméré, Caroline Earle, Julia Fahlke, Erich Fitzgerald, John Flynn, Andrew Foote, Ewan Fordyce, Ari Friedlander, John Gatesy, Denis Geraads, Phil Gingerich, Pavel Gol'din, Uko Gorter, Richard Hulbert, Olivier Lambert, Bill Monteleone, Phil Morin, Mizuki Murakami, Mary Parrish, Klaus Post, Nick Pyenson, Tim Scheirer, Eric Scott, Art Spiess, William Stout, Hans Thewissen, Doyle Trankina, Cheng-Hsiu Tsai, Ana M. Valenzuela-Toro, and Jorge Vélez-Juarbe. Ray Troll is acknowledged for his wonderfully inspired stylized cover art.

Many colleagues have provided valuable comments and suggestions that have helped shape the content of this book. I especially want to acknowledge the following for providing critical reviews and suggestions on drafts of various book chapters: Brian Beatty, Robert Boessenecker, Morgan Churchill, Tom Deméré, Eric Ekdale, Paul Koch, Hans Thewissen, Mark Uhen, and Jorge Vélez-Juarbe. The following institutions permitted the use of images under licensing agreements: Florida Museum of Natural History, the Smithsonian Institution, and the Ashoro Museum of Paleontology.

For providing funds toward publication I thank Stan Maloy, Dean of the College of Sciences, San Diego State University. I would also like to thank the editorial team at Johns Hopkins University Press, especially Mary Lou Kenney, Jen Malat, Juliana McCarthy, and Meagan M. Szekely, for their expert advice and assistance, and Linda Strange for meticulous and thorough copy editing. Editors Vincent Burke and Tiffany Gasbarrini are acknowledged for their patience and encouragement during preparation of this book.

The Rise of

Marine Mammals

1 Setting the Stage

Rocks, Fossils, and Evolution

Marine mammals have long captured the attention of humans. From early observations of seals and dolphins etched on the walls and ceilings of Paleolithic caves to the use of twenty-first-century satellites and microprocessors to track the underwater movements of these denizens of the deep, the lives of humans and marine mammals have been inexorably entangled. This connection has been a compelling story of the discovery and scientific study of large, charismatic sea animals, tempered by darker times of intense human hunting and whaling that left many species hovering on the brink of extinction. What is less well appreciated is that understanding the evolutionary history of marine mammals—past environmental challenges and biotic pressures (e.g., competition and predation) faced by these species and the anatomical and physiological adaptations enabling their survival—informs our understanding of the way that marine mammals respond to global climate change today and will respond in the future. Marine mammals are indicator species of the health of the ecosystems they inhabit and serve a vital role as sentinels of climate change. Perhaps most importantly, knowledge of the biology of marine mammals provides a framework that is essential for helping us understand how best to protect and conserve these animals. Study of fossil marine mammals, in particular, informs us about historical changes in past communities, providing a valuable context for predicting factors that will determine future vulnerability to extinction.

Marine mammals comprise a diverse assemblage of at least seven distinct evolutionary lineages that independently returned to the sea and spend the majority of their time in water. Today, marine mammals include some 125 living species worldwide. They occupy a variety of environments, ranging from shallow coastal waters and bays to deep open ocean, and live in freshwater, estuarine, and marine habitats. Marine mammals have been on Earth for a little more than 50 million years. Their evolution began on land, and their common ancestors had well-developed limbs and feet. The transition from land to sea involved different mammalian lineages, each converging on an aquatic lifestyle and acquiring many diverse adaptations, including those for locomotion, feeding, respiration, and hearing. These adaptations led to their evolutionary success as giants of the sea, occupying ecologically important roles in ocean food webs.

One aim of this book is to present some of the remarkable discoveries of fossil marine mammals, with particular reference to the origins, diversification, and phylogenetic relationships of living and extinct lineages. Fossils provide the only direct evidence of extinct species and past morphologies; they indicate extinct clades and ecologies not seen today. Fossils also provide evidence of behavior and habitats, allowing reconstructions of the modes of life of extinct marine mammals. A second objective is to integrate fossil discoveries in the context of the main events in Earth history that define the evolution of marine mammals. Changes in the positions of continents and sea margins have affected ocean circulation patterns and, consequently, food availability and marine mammal distributions. A final objective is to illustrate how the evolutionary biology of marine mammals has been enriched by recent remarkable advances in integrative and collaborative research that bring together paleontology, molecular biology, ecology, behavioral biology, genetics, and developmental biology. For example, exciting breakthroughs in understanding the anatomy and development of fetal whales have been facilitated by modern techniques such as CT scans and 3-D imagery (figure 1.1). Isotope studies reveal ocean temperature changes through time, providing evidence that the diversification of various marine mammal lineages was associated with increased food production, and providing information about the paleoecology of various marine species. Genetic and genome studies have explored evolutionary relationships among marine mammals and important transitions in the fossil record (e.g., hind-limb loss in cetaceans), as well as providing valuable life history and population data critical to the development of thorough management and conservation plans.

In this chapter, I present a brief introduction to marine mammal fossils—their naming, description, and organization in an evolutionary hierarchy, as well as methods of their discovery, collection, and preparation. Data from specimens in museum collections provide a historical

Figure 1.1. Photograph and 3-D reconstruction from CT images of a blue whale (*Balaenoptera musculus*) fetus. The specimen was collected in 1936 and is preserved in alcohol. United States National Museum 260581. Photo by M. Yamato, Smithsonian Institution.

context for interpreting present-day diversity and ecology and an understanding of how living species evolved. In recent decades, critical insights into the behavior of extant species have come from technological advances. For example, monitoring devices such as digital acoustic tags, developed in the 1990s, when attached to marine mammals have elucidated extraordinary feeding behaviors and foraging strategies (figure 1.2). Not only do these tags provide information on body orientation (acceleration, pitch, roll, and heading), they also record sounds made by and heard by the tagged marine mammals and record environmental parameters such as water temperature and depth (figure 1.3).

Naming, Describing, and Classifying

There are three major living groups of marine mammals. The most numerous, with 89 currently recognized extant species, is Cetacea (from the Greek *cetus*, meaning "whale"), which includes whales, dolphins, and porpoises. The second largest group of marine mammals, consisting of 36 extant species, is Pinnipedia (from the Latin *pinna*, meaning "fin," and *pes*, "foot"), "fin-footed" aquatic members of Carnivora: seals, sea lions, and wal-

Figure 1.2. Artist's rendition of a video and data recorder on a Weddell seal capturing prey. Photo courtesy of W. Monteleone.

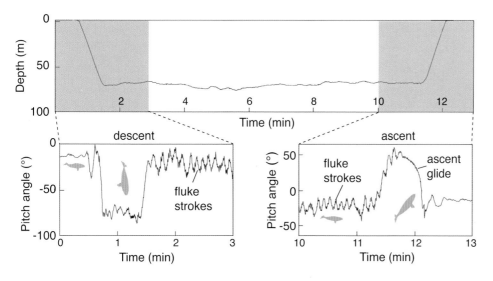

Figure 1.3. Swimming and dive data for a tagged right whale, *Eubalaena glacialis*. Modified from Nowacek et al., 2001.

ruses. The third major group, Sirenia, which derives its name from the legendary sirens of Greek mythology, contains four extant species: the dugong and three species of manatee. Other lineages of marine mammals that inhabit the sea and spend most of their lives in water are polar bears and sea otters, which are also carnivorans. Marine mammals were abundant at various times in the past, with some lineages more diverse than at present, in-

cluding extinct taxa such as the hippopotamus-like desmostylians, the bizarre carnivoran *Kolponomos*, the aquatic sloth *Thalassocnus*, and the recently extinct sea mink, *Neovison*. Figure 1.4 shows the various groups of living and some extinct marine mammals and their species-level diversity.

To communicate about past and present biodiversity, it is important to be able to define a species. Despite dis-

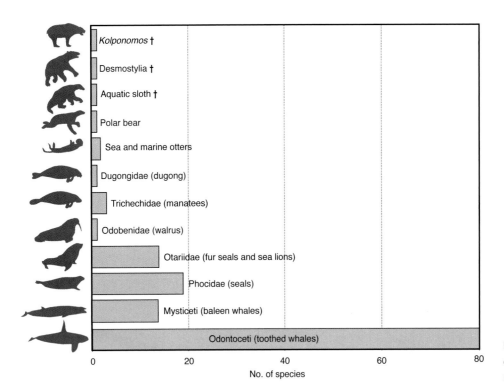

Figure 1.4. Diversity of major extant and extinct lineages of marine mammals.

agreements about what constitutes a species, it is widely understood that naming and identifying species is an important initial step. Ultimately, such decisions guide species protection efforts, since the number of living individuals within a species is a primary basis for designating conservation status. For example, fewer than 30 individuals of the vaquita are left, giving it the dubious distinction of being the most endangered marine mammal alive today. The Linnean system, named for the seventeenth-century Swedish botanist Carl von Linné (Linnaeus), is a binomial naming system comprising a genus name and a species name that have Latin or Greek roots, also referred to as the scientific name. For example, *Phocoena sinus* is the scientific name for the vaquita. Since scientific names are unique, they ensure that scientists, regardless of their native tongue, use the same universal language when communicating with one another. The same species may have different common names. For example, *Phocoena sinus* is also known as the gulf porpoise or desert porpoise, but biologists of all nationalities recognize *Phocoena sinus* as referring to a specific kind of porpoise.

Species can be organized into nested hierarchies based on the distribution of their shared derived characters (known as synapomorphies). Characters can be defined as heritable attributes of organisms that include anatomical features, DNA sequences, and behavioral traits. Characters that are inherited from a common ancestor are said to be homologous. A shared derived character is one that is different from the ancestral character (which does not indicate relatedness) and can thus be used to define a more inclusive group of species. For example, pinnipeds have certain features of the fore flipper such as an elongated first digit (thumb equivalent)—a synapomorphy—that unite pinnipeds and distinguish them from their close terrestrial arctoid relatives. Groups of species that can be shown to be united by shared derived characters are called monophyletic groups, or clades, indicating that they are descended from a common ancestor. A cladogram represents the temporal sequence of character acquisition—that is, hierarchies of shared derived characters within an evolutionary group known as a lineage, as illustrated for pinnipeds in figure 1.5. To infer evolutionary relationships within a group of species, also known as a phylogeny, one must determine which characters are derived and which are ancestral. If the ancestral condition of a character is established, then the direction of evolution, or polarity, from ancestral to derived can be inferred and synapomorphies recognized. Outgroup comparison is the most widely used procedure, which relies on the argument that a character state found in the outgroup (close relatives but outside the study group of interest) is likely to be the ancestral state for the clade in question (the ingroup), with derived states found in the ingroup. In figure 1.5, the ingroup is the major lineages of pinnipeds; the outgroup is all other arctoid carnivorans. Outgroups provide a point of comparison with the ingroup, making it possible for the tree to be rooted and to determine the direction (polarity) of character change. Next, the ingroup and outgroup are scored for the traits selected, and a character matrix is built.

Given the large size and complexity of data sets, modern cladistic analysis of living or extinct species, or both, uses computer programs to analyze characters and determine which species are closely related. When possible I have chosen to present various marine mammal phylogenies in terms of the results of combined morphological (anatomical) and molecular (genetic) data. Since fossils are the subject of this book, I have incorporated fossil species into phylogenies as well as using them to determine dates of divergence, in an effort to present an integrated view of the origin and diversification of

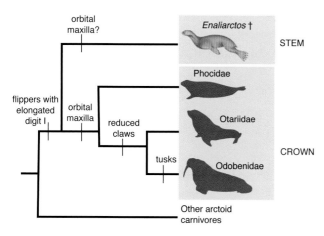

Figure 1.5. A cladogram depicting the evolutionary relationships among major lineages of pinnipeds. Modified from Berta et al., 2015.

various marine mammal lineages. However, often only morphology-based phylogenies exist for fossil taxa. Controversies remain regarding the phylogenetic position of various fossil taxa, and I discuss these in an effort to promote further research. I also explore the application of phylogenies to address broader evolutionary, ecological, and behavioral questions concerning marine mammals, such as the evolution of locomotion, body size, feeding, and hearing.

Important concepts when defining members of a clade are crown and stem groups. A crown group is the smallest clade to contain the last common ancestor of all extant members and all of that ancestor's descendants. For example, the crown group Pinnipedia is the clade of the last common ancestor of all living pinnipeds and fossil taxa, but not the stem lineage. The stem group includes taxa that fall close to but outside a particular crown group. For example, the fossil pinnipedimorph *Enaliarctos* can be classified with the stem group of pinnipeds, along with other extinct ancestral forms that are more closely related to living taxa than any other carnivoran (figure 1.5).

Only monophyletic groups accurately reflect evolutionary relationship; groups of species that lack shared characters are termed nonmonophyletic groups. Two types of nonmonophyletic groups can be distinguished, paraphyletic and polyphyletic. Paraphyletic groups include the most recent common ancestor and some but not all descendants of that ancestor. An example of a paraphyletic group is the "archaeocetes," an extinct group of cetaceans recognized as a taxon distinct from later-diverging whales, mysticetes (baleen whales), and odontocetes (toothed whales). Because some "archaeocetes" (e.g., basilosaurids) have been found to be closer to mysticetes and odontocetes than to other "archaeocetes," this group is rendered paraphyletic (figure 1.6, left). Polyphyletic groups, in contrast, are based on convergently evolved, nonhomologous characters. Inclusion of all river dolphins in a single group, as has historically been done, because they share characters that reflect adaptation to fresh water is an example of a polyphyletic group (figure 1.6, right). The current consensus based on both morphological and molecular data is that traits associated with

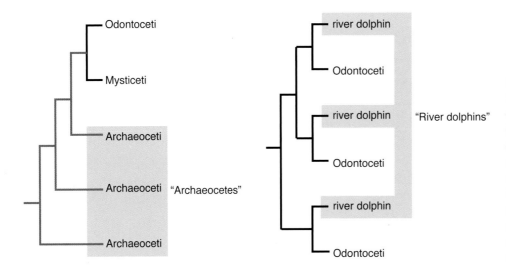

Figure 1.6. "Archaeoceti" is paraphyletic because it excludes some descendants (odontocete and mysticete) of the common ancestor of all "archaeocetes." "River dolphins" is polyphyletic because the river dolphins do not share a recent common ancestry and any anatomical similarities are independently derived.

Table 1.1 Comparison of Traditional Linnean and Cladistic Classifications

Linnean Classification		Cladistic Classification
Order	Carnivora	Carnivora
Suborder	Pinnipedia	Pinnipedimorpha
		Enaliarctos
Family	Otariidae	Otariidae
Genus and species	*Callorhinus ursinus*	*Callorhinus ursinus*

adaptation to fresh water evolved independently among different river dolphin lineages and that these shared characters among all river dolphins are therefore not homologous. Nonmonophyletic groups are not recognized by most systematists because these groups have the potential to misrepresent and distort evolutionary history.

Following the Linnean system of nomenclature, groups of organisms (e.g., species) are organized into higher categories or ranks (families, orders, classes, etc.). Given the arbitrariness of ranks above the species level—for example, a family of whales with only one species, such as gray whales (Eschrichtiidae), is not comparable to a family of pinnipeds with 19 species, such as true seals (Phocidae)—many biologists do not employ ranks above species (table 1.1). Thus, hierarchical levels in a cladistic classification based on common ancestry are conveyed by indentation in the listed names, rather than by named categories. Extant and closely related extinct species are

referred to as the crown group, and extinct fossil species at the base of the tree that preceded the point at which the crown group branched off (i.e., the most recent common ancestor of all extant species) as the stem group.

Cladistic classification reflects phylogeny and can be used to reconstruct the cladogram. Chapters 2 to 6, covering the evolution and paleobiology of extant and fossil marine mammal lineages, are organized based on cladistic phylogenetic frameworks obtained with reference to the best available hypotheses of evolutionary relationship that reflect the most recent research (see also Classification of Fossil Marine Mammals at the end of the book).

Terrestrial Affinities

Marine mammals have common ancestors that mostly lived on land. In fact, one of the most fascinating evolutionary transitions made by marine mammals was their secondary adaptation to aquatic lifestyles, including, for some lineages, dramatic changes in body size and shape. A brief review of the terrestrial affinities of the major lineages provides a framework for understanding the broad context of their separate evolutionary histories. Confirmation of the significance of the land mammal legacy of whales was noted by the University of Chicago evolutionary biologist Neil Shubin (2001), who described whale evolution as "tinkering with land mammals . . . using the old to make the new."

The phylogenetic position of whales among mammals has been much debated. Earlier workers argued for their affinities among both marine reptiles and various mammalian groups, including marsupials, insectivores, extinct creodonts, pinnipeds, edentates, artiodactyls, and perissodactyls. Fossil discoveries, especially in the past few decades, have revealed that whales are nested within Cetartiodactyla, a clade that also includes even-toed ungulates (artiodactyls) such as hippopotamuses, giraffes, and deer. Traditionally, an extinct group of land-dwelling ungulates known as mesonychid condylarths were thought to be the closest fossil relatives of whales. However, it is now generally accepted that extinct raoellid artiodactyls are the closest whale relatives (figure 1.7). Raoellids lived between 52 and 46 million years ago, and fossils are known from South Asia, Pakistan, and India. Although a key fossil locality in India, Kalakot, has not been studied sedimentologically, and not much is known about the habitat these animals occupied, it is possible, based on the hundreds of skeletons found at this site, that it was a floodplain of a river where animals lived and died. Several genera have been described, including *Indohyus*, *Khirtharia*, *Kunmunella*, and *Metkatius*. *Indohyus major*, the best-known species, resembles a tiny but more heavily built deer with a long tail (figure 1.8). *Indohyus* had a long, pointed snout with anterior teeth (incisors) arranged front to back rather than the more typical side to side as in most mammals, which suggests the teeth may have functioned as a mechanism for cropping plants. The bones and teeth of fossil marine mammals, based on concentrations of various isotopes (e.g., nitrogen, oxygen, and hydrogen), provide information about the food and

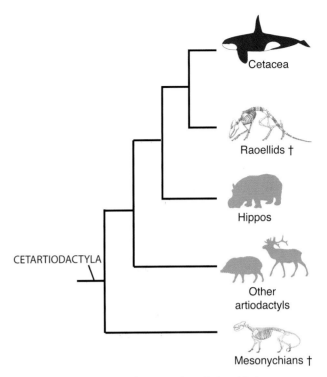

Figure 1.7. A cladogram depicting the relationship between cetaceans and their terrestrial relatives. Modified from Thewissen and Bajpai, 2009.

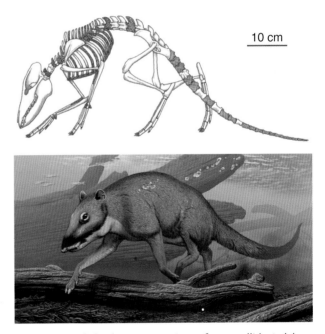

Figure 1.8. Skeletal reconstruction of a raoellid, *Indohyus*. *Hatched elements* are reconstructed on the basis of related taxa and life restoration. Reconstruction from Thewissen et al., 2007. Life restoration illustrated by C. Buell.

water they ingested. In this case, isotopic data confirm that *Indohyus* and stem artiodactyls had a diet of terrestrial plants.

The long, slender limbs of *Indohyus* possessed five fingers and four or five toes. *Indohyus* was digitigrade, walking on its digits like a dog, rather than on the tips of its toes like an ungulate. The dense limb bones of raoellids indicate that they could walk underwater, which suggests that an aquatic lifestyle arose before whales evolved. The upper ankle bone of raoellids consisted of a "double pulley" similar to that seen in other artiodactyls and extinct cetaceans. This shape of the ankle bone allowed greater anteroposterior mobility in the foot. Raoellids and the earliest cetaceans were probably able to wade and swim effectively, as their skeletal morphologies indicate. Additionally, their ears show specializations that suggest they could hear underwater sound. Whale paleontologist J. G. M. "Hans" Thewissen (2015) proposed an evolutionary scenario for cetaceans in which their terrestrial artiodactyl ancestors first took to the water for shelter from predators, as do some extant tragulids (mouse deer). From this initial step, early cetaceans spent successively more time in water, much like the modern hippopotamus, and eventually began feeding in water and transitioning from herbivory to carnivory/piscivory.

A novel study by biologist Scott Mirceta and colleagues in 2013 that combined data on body mass with inferred concentration of myoglobin, an oxygen-binding molecule in muscle cells, provides insight into the diving capabilities of extinct and living marine mammals. Based on estimates of fossil body mass, these scientists were able to estimate the dive time of *Indohyus* as approximately 1.6 minutes, similar to that for the hippopotamus, suggesting that the closest terrestrial relatives of whales were not capable of long, sustained dives. This study illustrates an important point that reconstruction of the evolutionary history of a molecule from living species can be used to predict the behavior of extinct species.

The past debate about the relationship of aquatic carnivorans, the pinnipeds (or pinnipedimorphs, to include their extinct stem relatives), to other carnivorans is now settled. The traditional view, diphyly, proposed that pinnipeds originated from two different carnivoran lineages: otariids (fur seals and sea lions) and odobenids (walruses) from ursids (bears), and phocids (seals) from mustelids (e.g., otters, weasels, badgers) (figure 1.9b). Morphological and molecular data strongly support a single origin, monophyly, for pinnipeds from arctoid carnivorans, although there is disagreement and acknowledged difficulty in sorting out whether ursids or musteloids are their closest terrestrial allies (figure 1.9a). The split between pinnipeds and ursids is estimated to have occurred approximately 35.7 million years ago, and that between pinnipeds and musteloids slightly later, 29.95 million years ago, based on molecular data that are broadly consistent with the fossil record.

The closest relatives of sirenians are proboscideans (elephants), hyracoids (hyraxes), and extinct desmostylians, recognized collectively as the clade Paenungulata (figure 1.10), according to both morphological and molecular phylogenies. A larger clade, Paenungulata + Afroinsectiphilia (aardvarks, tenrecs, golden moles, elephant shrews), unites these groups as Afrotheria, named for the African origin of its members. Afrotherians are known in the fossil record beginning approximately 60 million years ago. Molecular data indicate that sirenians diverged from paenungulates approximately 65 million years ago, which is considerably earlier than the earliest fossil record of sirenians, at approximately 52 million years ago.

Discovery, Collection, and Preparation

Fossils of marine mammals are the only direct record of their ancient past. The most well-preserved elements are hard parts such as bones and teeth. The discovery and collection of fossils require considerable time, effort, and luck. Localities for collecting marine mammal fossils are known worldwide, from the tropics to the poles and on every major continent. Usually, the fossils are found in sedimentary rocks composed of preexisting sediments such as sand, silt, or mud that may be cemented together by chemical solutions, such as calcium carbonate, that harden them into rocks. Sedimentary rocks are exposed on land as the result of crustal uplift, weathering, and erosion, although some records of marine mammal fos-

(a) Monophyly

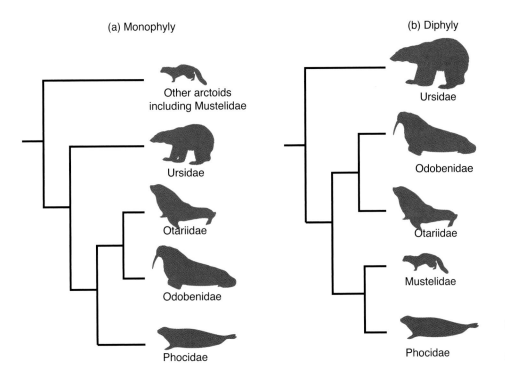

Other arctoids
including Mustelidae

Ursidae

Otariidae

Odobenidae

Phocidae

(b) Diphyly

Ursidae

Odobenidae

Otariidae

Mustelidae

Phocidae

Figure 1.9. Alternative hypotheses for relationships among pinnipeds.

sils come from deep ocean dredgings (see chapter 3). Discoveries range from small, localized sites (figure 1.11) to scattered occurrences of skeletal elements across many kilometers (figure 1.12). Sediments enclosing the bones and the bones themselves (i.e., isotope values) reveal the environment of deposition, such as whether the marine mammal lived on land or in marine, brackish, or fresh water, and in some cases reveal what caused the animal's death. For example, there may be evidence of predation on whale carcasses by bone-eating worms (a process known as bioerosion, also documented in modern whale fall communities; see chapter 7), or various encrusting invertebrates such as barnacles, sponges, and bivalves that use the bones for support/shelter, or shark feeding traces. Such information can be used in paleoenvironmental reconstructions.

Once the fossilized bones are exposed, the rock around them is scraped away during excavation by paleontologists. Excavation continues until the fossil and matrix are fully exposed, with some rock matrix left for support (figures 1.11, 1.12). The next step is to build a plaster jacket around the fossil to protect it from the elements

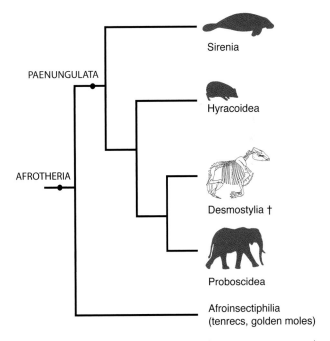

PAENUNGULATA

Sirenia

Hyracoidea

Desmostylia †

AFROTHERIA

Proboscidea

Afroinsectiphilia
(tenrecs, golden moles)

Figure 1.10. Evolutionary relationships between sirenians and their relatives. From Kuntner et al., 2010.

Figure 1.11. Field preparation of the fossil sperm whale *Livyatan* at Cerro Colorado, Peru. Courtesy of G. Bianucci.

Figure 1.12. Smithsonian crew excavating fossil whales in Chile. Courtesy of Nick Pyenson and the Smithsonian Institution.

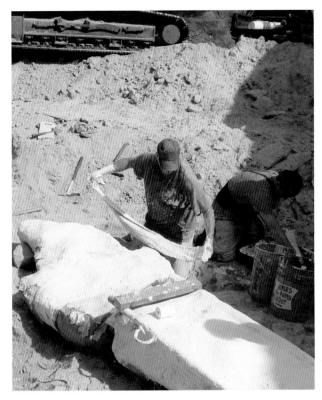

Figure 1.13. Crew from the San Diego Natural History Museum building a plaster jacket for a fossil baleen whale. Courtesy of T. Deméré.

and during transport to a museum for preparation and storage (figure 1.13).

When the fossil reaches the laboratory, preparation begins (figure 1.14). With an array of mechanical tools including air scribes and abrasives, the rock around the fossil is carefully removed to reveal the bones. Today, advances in fossil preparation also include noninvasive techniques such as micro CT and 3-D laser scanning. For example, in Chile, when highway construction threatened to destroy a spectacular collection of fossil baleen whales, Smithsonian scientists and staff used 3-D surface laser scans to document the arrangement of the skeletons before their removal (Pyenson et al., 2014) (figure 1.15).

The final step in the preparation process is curation of the fossil into a public institution so that it is available for exhibit or study. Curation involves identifying and labeling the specimen, linking information about the specimen and associated data (species, locality, collector), and storing the fossil to promote its long-term preserva-

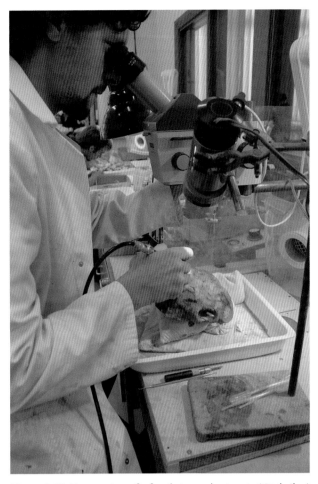

Figure 1.14. Preparation of a fossil stem odontocete (*Otekaikea*) by former PhD student Gabriel Aguirre-Fernández in New Zealand. Courtesy of R. E. Fordyce.

tion. Some specimens are prepared for public exhibition, whereas others are stored in the museum's collection facilities. Given the fragile and brittle nature of fossil bones and the need to make them available for study, mounted specimens are often casts of bones in fiberglass or other resins that show lifelike dynamic poses (figure 1.16).

Collections professionals at natural history museums have the responsibility of preserving, cataloguing, organizing, and storing fossils and other specimens of our natural world. Specimens put on display encourage public appreciation and inspiration, while fossils stored in collections are available for scientific study, today and in the future (figure 1.17). Depositing vouchered specimens—those representative of a taxon, along with details on locality and other attendant data—in accessible

Figure 1.15. Laser scanning of a baleen whale in Chile by Smithsonian staff. Courtesy of Nick Pyenson and the Smithsonian Institution.

Figure 1.16. *Allodesmus* from Sharktooth Hill in a dynamic mount at San Diego Natural History Museum. Courtesy of E. Scott.

natural history collections is essential to confirm specimen identity and prevent misidentification in future studies. This is especially important for rare or extinct taxa. Collections-based research ensures that fossils serve a variety of purposes, including providing knowledge about the diversity and evolution of various organismal groups and patterns of morphological change over time. The growing availability of digital technology and internet access (e.g., online availability of images in the absence of physical specimens) offers new ways to create active learning opportunities through the study of fossil and other natural history specimens.

The Geologic Time Scale, Tectonics, and Marine Mammals

The geologic time scale divides Earth's history into time intervals ranging from oldest to youngest and provides absolute measurements of their duration. The relative time scale has named intervals (epochs; figure 1.18), and as far as marine mammals are concerned, our interest lies within the Cenozoic Era, which is divided into the Paleocene, Eocene, Oligocene, Miocene, Pliocene, and Pleisto-

Figure 1.17. Tom Deméré, paleontology curator, with marine mammal fossils, San Diego Natural History Museum.

cene epochs. These epochs are usually divided into early, middle, and late intervals. Geologic stages (e.g., Langhian) and Land Mammal Ages (e.g., Irvingtonian) provide even finer subdivision. Absolute dates are determined by measuring the radioactive decay of certain isotopes preserved in rocks (figure 1.18). Because radioactive isotopes break down, or decay, at a predictable rate (characterized by half-life), by determining the rate of decay recorded in the mineral or rock one can infer the age of the rock and the fossil within it. The evolution of marine mammals is intimately tied to Earth's history.

From the theory of plate tectonics we know that the earth's surface is divided into a series of plates consisting of large chunks of the earth's crust that have collided and separated in the geologic past, forming or enlarging existing oceans and rearranging the world's continents— which carry biologically diverse faunas, including marine mammals. As discussed in later chapters, geologic and climatic events such as the rise and fall of sea levels, tectonic events, and ocean circulation patterns have also affected the origin and diversification of marine mammals (figure 1.19). One of the earliest major geologic events affecting marine mammal diversification was the separation of Antarctica and South America in the late Oligocene (30-23 million years ago). This restructuring allowed the Antarctic icecap to expand and global climates to cool, which resulted in increased zooplankton productivity and is hypothesized to have led to the explosive radiation of both mysticete and odontocete cetaceans. Another event, opening of the Central American (Panamanian) seaway separating North and South America, allowed the exchange of waters and biota (until 11 million years ago; it then opened again from 6 to 4 million years ago) and served as a dispersal route for monachine seals, odobenine walruses, and dugongid sirenians. The Bering Strait, a seaway between Alaska and Siberia, which opened as the result of tectonic activity during the late Miocene to earliest Pliocene (5.3-4.8 million years ago), was an important dispersal route followed by the extant walrus, bowhead whale, several phocine seals, hydrodamaline dugongids, and the gray whale. During the Pleistocene (1.6 million years ago), glacial and interglacial oscillations reduced gene flow between isolated populations and probably drove speciation in some Arctic phocine seals.

Diversity through Time: A Rock Bias?

Paleontologists have long noted a close connection between the rock record and the fossil record. This is based on the assumption that time intervals represented by thick accumulations of fossil-containing sedimentary rocks allow larger numbers of fossils to be collected per sampled time interval. Alternatively, it is possible that both sediment abundance and species diversity are driven by other factors such as climate change or competition and replacement. Vertebrate paleontologist Felix Marx (2009) examined the diversity of major marine mammal groups through time—cetaceans, pinnipedimorphs, and sirenians—based on the western European rock record and found a general similarity in the diversity curves of cetaceans and pinnipedimorphs, with a diversity peak during the middle Miocene (Burdigalian) followed by an apparent crash during the late Miocene (Messinian) (figure

Geologic and climatic events

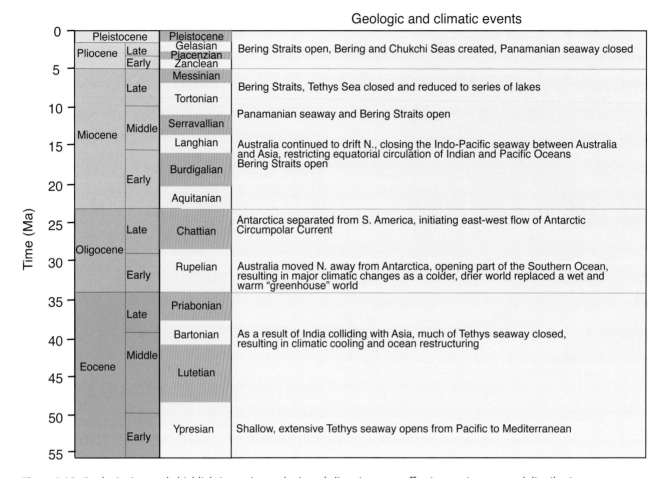

Time (Ma)				Geologic and climatic events
Pleistocene			Pleistocene	Bering Straits open, Bering and Chukchi Seas created, Panamanian seaway closed
Pliocene	Late		Gelasian	
			Piacenzian	
	Early		Zanclean	
Miocene	Late		Messinian	Bering Straits, Tethys Sea closed and reduced to series of lakes
			Tortonian	
	Middle		Serravallian	Panamanian seaway and Bering Straits open
			Langhian	Australia continued to drift N., closing the Indo-Pacific seaway between Australia and Asia, restricting equatorial circulation of Indian and Pacific Oceans
	Early		Burdigalian	Bering Straits open
			Aquitanian	
Oligocene	Late		Chattian	Antarctica separated from S. America, initiating east-west flow of Antarctic Circumpolar Current
	Early		Rupelian	Australia moved N. away from Antarctica, opening part of the Southern Ocean, resulting in major climatic changes as a colder, drier world replaced a wet and warm "greenhouse" world
Eocene	Late		Priabonian	
	Middle		Bartonian	As a result of India colliding with Asia, much of Tethys seaway closed, resulting in climatic cooling and ocean restructuring
			Lutetian	
	Early		Ypresian	Shallow, extensive Tethys seaway opens from Pacific to Mediterranean

Figure 1.18. Geologic time scale highlighting major geologic and climatic events affecting marine mammal distributions.

1.20). The cetacean/pinnipedimorph pattern contrasted with a sirenian peak during the early Miocene followed by rapid decline and continuous low levels of diversity from the mid-Miocene onward.

Results of Marx's study corroborate those of paleontologists Mark Uhen and Nick Pyenson (2007), who failed to find a strong association between rock outcrop area and cetacean paleodiversity in North America. Given that these results contrast with those for marine invertebrate Phanerozoic (covering the whole of time since the beginning of the Cambrian period) diversity, other explanations for this discrepancy are possible. For example, examination of data on marine mammal occurrences downloaded from the Paleobiology Database (https://paleobiodb.org) in 2007 revealed that of 129 cetacean genera, nearly two-thirds were known by fewer than five occurrences, with

40% known by a single occurrence. The figures were similar for pinnipedimorphs (62% fewer than five occurrences; 31% single occurrences) and sirenians (80% fewer than five occurrences; 33% single occurrences). In downloads from the Paleobiology Database as of September 19, 2016, revealing an almost doubling of the number of cetacean genera, similar occurrences were noted, with more than two-thirds (nearly 70%) known by fewer than five occurrences and 34% by single occurrences. The numbers were similar for pinnipedimorphs and sirenians and were lower than those for cetaceans (50% fewer than five occurrences; 20% single occurrences). Given these low abundances, high preservation potential seems unlikely. An alternative explanation may be environmental or biological change, or both, and the fact that closely related taxa with a similar biology will respond in a similar way

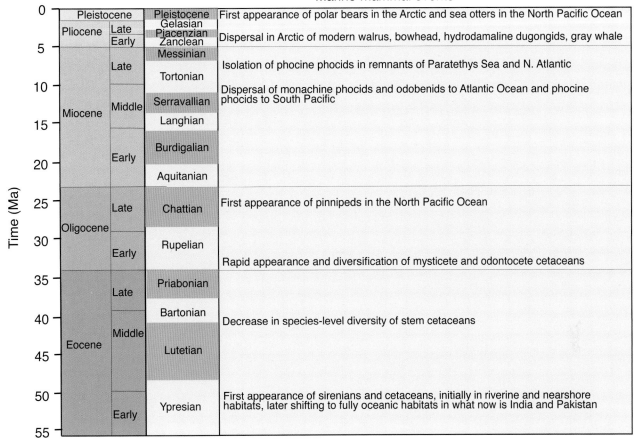

Figure 1.19. Geologic time scale and major events in marine mammal evolution.

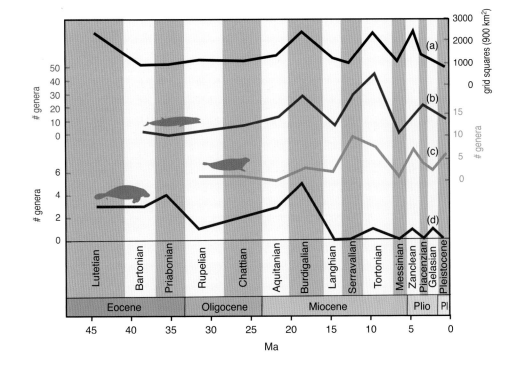

Figure 1.20. Plots of raw data on European marine sediment outcrops, attributed to geologic stages according to their relative durations and numbers, as shown in plot (*a*). Below are plots of (*b*) cetacean, (*c*) pinnipedimorph, and (*d*) sirenian taxon counts. From Marx, 2009.

to such change. By contrast, groups with widely different biologies are more likely to show different responses to a given environmental change. Viewed in this way, cetacean and pinnipedimorph diversity curves may reflect a similar response to external influences such as changing sea levels, whereas sirenian paleodiversity curves differ in part due to a difference in diet (herbivorous vs. carnivorous) and habitat (shallow coastal areas near the tropics vs. offshore and open waters). As discussed further in chapter 7, both environmental and biological factors need to be explored and tested for their influence on marine mammal paleodiversity.

2 The Oldest Marine Mammals

Whales and Sea Cows

When marine mammals first evolved, beginning in the early to late Eocene (54–34 million years ago), an extensive, shallow Tethys Sea (named for a sea goddess of Greek mythology) stretched from the Pacific to the present-day Mediterranean. Stem whales originated and diversified on the eastern shores of the Tethys Sea (in what is now India and Pakistan). Stem sirenians also had an Old World origin, first appearing in northern and western Africa and dispersing shortly thereafter via the Atlantic to the Caribbean (figure 2.1). Among the best-known fossil localities is the middle to late Eocene (48–37 million years ago) Fayum basin in Egypt, including the Valley of Whales, which has yielded a considerable diversity of fossil whales and sirenians. Paleontologist Phil Gingerich (1992) has interpreted the fossiliferous strata in terms of changing sea levels, ranging from deposition offshore in marine shallow shelf environments to deposition in nearshore and lagoonal environments. The climate at that time was much warmer than today, with tropical faunas at latitudes now at the Arctic Circle. This initial radiation of stem sirenians and cetaceans during the middle Eocene involved shifts from riverine to nearshore coastal settings and then to open ocean habitats, and both marine mammal groups became widely distributed geographically.

The transition from a terrestrial ancestry to a fully aquatic life in both cetaceans and sirenians involved a wide range of morphological and physiological adaptations, including feeding, locomotion, respiration, and hearing. Fossils, along with new techniques for studying them (e.g., CT and 3-D reconstructions) that combine molecular, physiological, and evolutionary approaches, have provided fresh insights into our understanding of the evolution of these adaptations. Much of our knowledge of stem whales comes from the impressive collection of fossil whales in Pakistan and India, beginning in the 1990s, and their anatomical study by whale paleontologist Hans Thewissen and his colleagues, as recounted in the book *The Walking Whales* (Thewissen, 2015).

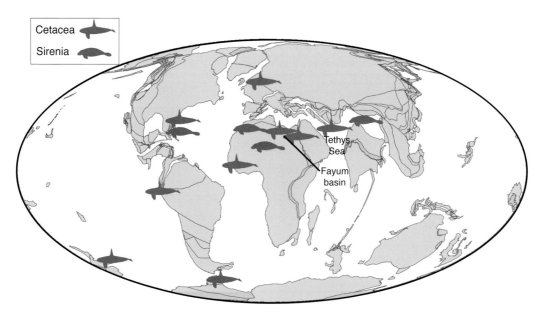

Figure 2.1. Major localities of marine mammal fossils, early to late Eocene (54–34 million years ago). Modified from Fordyce, 2009; Benoit et al., 2013. Base tectonic map from www.odsn.de/odsn/about.html.

From Walking to Semiaquatic Stem Whales

The English naturalist John Ray (1693) regarded whales as fish, but Aristotle had been the first to recognize that whales shared many characters with mammals, and identified these animals as Cete or Cetacea (see Gingerich, 2015b). The oldest fossil whales, commonly referred to as "archaeocetes," are distinguished from living whales on the basis of ancestral characters, including nostrils positioned near the tip of the nose, with some "archaeocetes" retaining well-developed hind limbs and others showing limb reduction and loss of attachment of the pelvis and hind limb to the vertebrae. Based on systematic work, four families of stem whales are recognized: Pakicetidae, Ambulocetidae, Protocetidae, and Remingtonocetidae. These taxa date back to the early and middle Eocene (50 million years ago) in Pakistan and India (figure 2.2).

Exploring Pakistan's early Eocene formations in the 1970s, vertebrate paleontologist Robert "Mac" West (1980) was the first to recognize whales in Pakistan, and a year later Gingerich found a skull that appeared to be a mesonychid condylarth but had ears like a whale's. Gingerich and Russell (1981) named the fossil *Pakicetus*, meaning "Pakistani whale." *Pakicetus* and its kin had wolf-sized (*Pakicetus, Nalacetus*) to fox-sized (*Ichthyolestes*) proportions, with long noses and tails (figures 2.3, 2.4). The eyes of pakicetids were close together and faced upward (dorsally), and the closely set eyes made the area between the eyes very small. Although the nose was long, the sense of smell was limited. The nose opening is perforated by small holes through which nerves probably traveled, indicating that pakicetids had a sensitive snout with vibrissae (whiskers), perhaps specialized for detecting motion in water, as in modern pinnipeds. The teeth of pakicetids possess sharp, shearing blades (figure 2.5), and wear on the teeth is consistent with a fish-eating habit. Tooth isotope values are consistent with pakicetids feeding on freshwater aquatic prey (see figure 2.6).

Discoveries of pakicetid skeletons indicate that they had well-developed fore and hind limbs (figures 2.3, 2.4). The postcrania of pakicetids resemble those of artiodactyls, exhibiting skeletal adaptations (such as long, slender digits) linked to cursoriality (running). Lumbar vertebrae of pakicetids display well-developed transverse and ac-

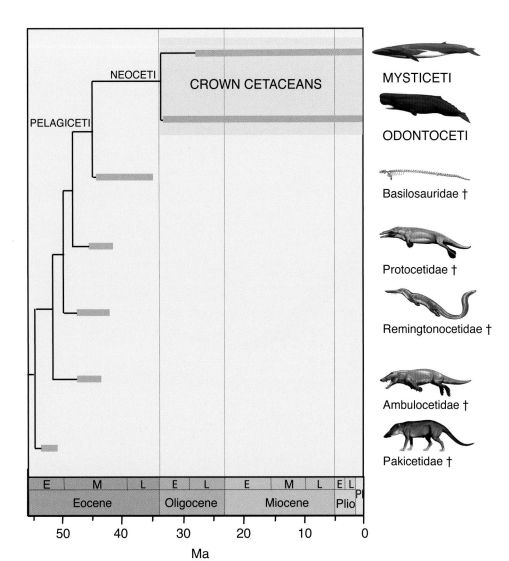

MYSTICETI

ODONTOCETI

Basilosauridae †

Protocetidae †

Remingtonocetidae †

Ambulocetidae †

Pakicetidae †

Figure 2.2. Phylogeny of stem whales. From Gatesy et al., 2013.

cessory processes, indicating increased surface area for powerful epaxial (trunk) musculature. Features of the vertebral column are consistent with use of the tail in dorsoventral movements during swimming. Elongated foot elements with muscular crests on the digits suggest webbed feet that could increase stability on wet, muddy surfaces. Isotopic analyses indicate that pakicetids were mostly freshwater animals, and they are believed to have waded and walked in freshwater streams (figure 2.6).

Ambulocetids include three genera: *Ambulocetus, Gandakasia*, and *Himalayacetus*, from the early to middle Eocene of India and Pakistan. *Gandakasia* was the first am-

bulocetid to be described, in 1958, and is known only by a few teeth. *Himalayacetus*, based on a lower jaw, was named for its discovery in the Indian Himalayas and was originally described as a pakicetid; it was thought to be the oldest whale, at 53.5 million years ago. It now seems that dating was based on associated fossils that washed in from older layers. Based on later systematic work, *Himalayacetus* was assigned to the ambulocetids, all of which lived approximately 48 million years ago.

Ambulocetids are best known by *Ambulocetus natans*, also known as the walking, swimming whale, described by Thewissen and colleagues (1994). It looked like a croc-

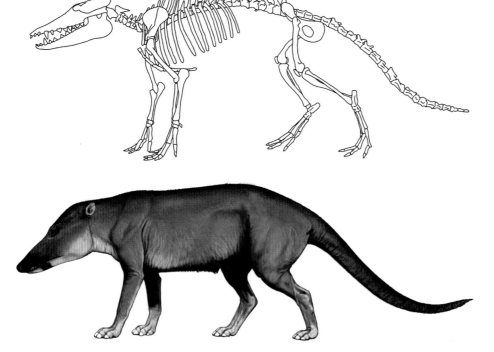

Figure 2.3. Skeleton of *Pakicetus*. From Thewissen, 2015.

Figure 2.4. Life restoration of *Pakicetus*. Illustrated by C. Buell.

odile, with a long-snouted skull, short forelimbs, and a powerful tail (figures 2.7, 2.8). Ambulocetids were much larger than pakicetids—approximately the size of a large male sea lion. The teeth and wear surfaces of *Ambulocetus* are similar to those in other stem whales. Tooth isotope values indicate that ambulocetids fed on freshwater aquatic vertebrate prey (figure 2.6), although the specimen was found in rocks formed in a shallow marine environment. The limb proportions of *A. natans* indicate that its shorter hind limbs and broad feet enabled it to swim, much like otters, with its hind legs and tail. Its powerful limbs suggest that, like crocodiles, it was an ambush predator, pursuing prey on land or in water.

Remingtonocetids are known by six genera: *Andrewsiphius, Attockicetus, Dalanistes, Kutchicetus, Rayanistes,* and *Remingtonocetus* from the middle Eocene of Pakistan and India (figure 2.9). They are usually regarded as sister group of the clade that includes Protocetidae, Basilosauridae, and Neoceti (figure 2.2). Remingtonocetids exhibit considerable diversity in size, ranging from the smallest, the sea-otter-sized *Kutchicetus* (figure 2.10), to the larg-

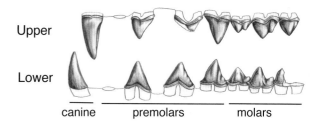

Figure 2.5. Dentition of the upper and lower jaws of *Pakicetus*. From Thewissen, 2015.

est, *Dalanistes*, weighing as much as a male sea lion. They were crocodile-like animals characterized by long, narrow skulls (as much as six times longer than wide) and jaws, tiny eyes, and big ears. Cranially, *Remingtonocetus* is the best known. The ear of remingtonocetids is specialized, as is the development of a large opening in the lower jaw, the mandibular foramen, which has an important function in hearing. *Ambulocetus* and *Himalayacetus* have a mandibular foramen that is intermediate in size between that of land mammals and that of basilosaurids and odontocetes. In crown whales (especially extant odontocetes),

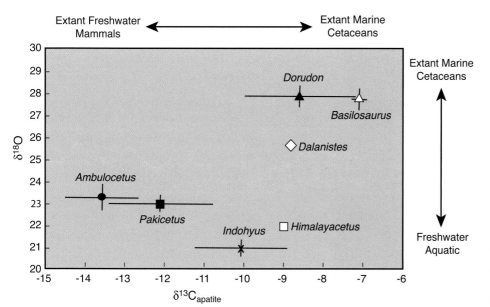

Figure 2.6. Mean isotopic carbon ($\delta^{13}C$) and oxygen ($\delta^{18}O$) values for fossil cetaceans and their relatives (*Indohyus*). Modified from Newsome et al., 2010.

the large mandible foramen contains a pad of fat that connects the lower jaw to the middle ear and provides a pathway for transmission of underwater sounds. Remingtonocetids' widely spaced molars that lack crushing basins show that they lost the crushing specialization of pakicetids and ambulocetids, which suggests a difference in diet from earlier cetaceans. The tall, large teeth of remingtonocetids probably functioned to capture and stabilize prey. Study of the feeding apparatus of remingtonocetids by Lisa Cooper and colleagues (2014) (e.g., the long mandible, tooth row, and mandibular symphysis and the narrow palatal arch) suggests that they engaged in snap feeding, as seen in river dolphins (figure 2.11). On further study, the long snout of remingtonocetids (e.g., *Andrewsiphius, Kutchicetus*) with unusual bone textures at the tip of the rostrum may reveal the presence of vibrissae. Unlike ambulocetids, remingtonocetids (e.g., *Dalanistes*) are also found in nearshore coastal settings, indicating a greater reliance on marine prey, which is consistent with tooth isotope values (figure 2.6). Study of a partial skeleton of *Remingtonocetus domandaensis* by paleontologist Ryan Bebej and colleagues (2012) suggests that this whale had robust limbs with some weight-bearing ability and that it swam by powerful movements of its hind limbs rather than undulations of its posterior vertebral column.

Rayanistes afer, the only remingtonocetid from the middle Eocene of Egypt, extends the geographic range of remingtonocetids to Africa (Bebej et al., 2016). It is distinguished from closely related taxa in having a specialized mode of locomotion. The expanded ischium and supporting musculature indicate that *Rayanistes* was capable of powerful hind-limb retraction during swimming. Additionally, lumbar vertebrae with posteriorly inclined neural spines are suggested to have increased the length of the power stroke during pelvic paddling. Recovery of *R. afer* in Africa demonstrates that remingtonocetids were skilled swimmers having the locomotor capability to cross the southern Tethys Sea between India-Pakistan and Africa; more likely, they swam across the island chain now occupied by Iran, Iraq, and the Arabian peninsula.

Although protocetids originated in India and Pakistan, the occurrences of later-diverging semiaquatic protocetid whales in Asia, Africa, Europe, and North and South America indicate that cetaceans crossed ocean basins and spread across the globe between 49 and 40 million years ago. The family Protocetidae was originally named by Stromer in 1908 after the genus *Protocetus*, based on a specimen that consisted of a skull and much of a skeleton from the Mokattam Formation in Egypt. Protocetids are the most numerous, diverse, and well-represented of

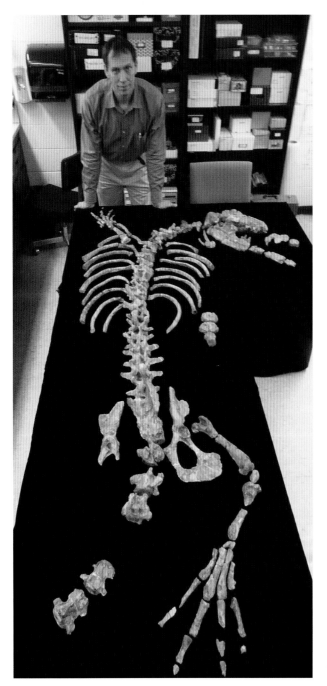

Figure 2.7. Vertebrate paleontologist Hans Thewissen and *Ambulocetus* skeleton in lab.

stem cetaceans (currently 20 genera are known). Protocetids from India and Pakistan are *Artiocetus, Babiacetus, Dhedacetus, Gaviacetus, Indocetus, Kharodacetus, Maiacetus, Makaracetus, Qaisracetus, Rodhocetus,* and *Takracetus.* African protocetids are *Aegyptocetus, Pappocetus, Protoce-*

tus, and *Togocetus.* North American protocetids are *Carolinacetus, Crenatocetus, Georgiacetus,* and *Nachitochia.* They differed from other early cetaceans in having large eyes and a nasal opening that had migrated further posteriorly on the skull. Their powerful, heavy jaws and large teeth suggest a diet composed of fish or other vertebrates. Tooth isotope values support a more fully marine lifestyle. Although the teeth of most protocetids are slender and high-crowned, the blunt, worn, and in some cases broken teeth of some taxa such as *Babiacetus* suggest a diet that included large, struggling prey such as marine catfish. Notable among protocetids is *Kharodacetus,* one of the largest known protocetids, similar in size to the basilosaurid *Zygorhiza.*

One of the more unusual discoveries of a fossil whale is that of the protocetid *Aegyptocetus tarfa,* uncovered by quarry workers in a block of limestone from the middle Eocene of Egypt. The exceptional preservation of the specimen in slabs showing cross sections reveals well-developed ethmoidal turbinal bones, indicating retention of a functional sense of smell (figure 2.12). Ventral deflection of the rostrum and frontal portions of the skull (clinorhynchy) suggests a rare bottom-feeding specialization or, more likely, is related to hearing and the localization of sound.

Structures associated with a sense of smell have been described in several other stem whales, including a remingtonocetid, a pakicetid (*Ichthyolestes*), as well as a skull provisionally referred to a protocetid (Godfrey et al., 2012). These structures, which include ethmoturbinates, olfactory meatus, and chambers for olfactory tract and olfactory bulb, provide further evidence of the importance of olfaction early in cetacean evolution. Olfactory organs were present in stem cetaceans and probably began to lose their function in remingtonocetids (figure 2.13). Mysticetes have been increasingly recognized as possessing functional olfactory systems, and they appear to have retained a sense of smell, which in turn may have allowed them to detect odors given off by plankton (krill) prey. Although some stem odontocetes (e.g., *Squalodon*) had relatively large olfactory bulb chambers and well-developed cribriform plates, crown odontocetes reduced

Figure 2.8. Life restoration of *Ambulocetus natans*. Illustrated by C. Buell.

Figure 2.9. Hans Thewissen in the field, sorting *Andrewsiphius* remains. Courtesy of J. G. M. Thewissen.

Figure 2.10. Life restoration of *Kutchicetus*. Illustrated by C. Buell.

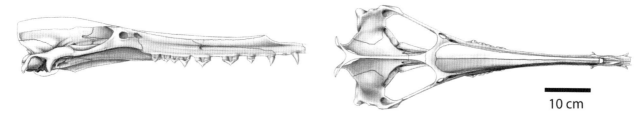

10 cm

Figure 2.11. Skull and jaws of a remingtonocetid. Courtesy of J. G. M. Thewissen.

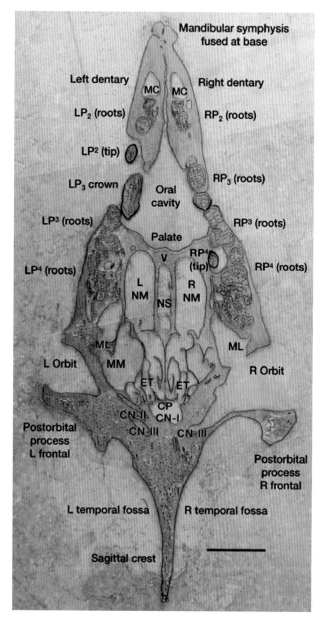

Figure 2.12. Cross section of skull of *Aegyptocetus tarfa*, showing turbinates. *Abbreviations:* CN-I, cranial nerve I (olfactory); CN-II, cranial nerve II (optic); CN-III, cranial nerve III (oculomotor); CP, cribriform plate at anterior end of olfactory tract (cranial nerve I); ET, ethmoidal turbinates in the caudal frontal sinus; MC, mandibular canal; ML, lateral maxillary sinus; MM, medial maxillary sinus; NM, nasopharyngeal meatus; NS, nasal septum with a cartilaginous core. Scale bar = 5 cm. From Bianucci and Gingerich, 2011.

their chemosensory abilities further, losing the entire olfactory bulb and most olfactory receptors. In the land-water transition of cetaceans, concomitant with changes in olfaction was a reduction in another chemical sense, that of taste, as suggested by genetic data (figure 2.13; see also chapter 3, figure 3.44).

Pelagiceti: Fully Aquatic Stem Whales

A later-diverging clade of whales, the Pelagiceti, includes the extinct Basilosauridae + Neoceti (crown cetaceans; figure 2.2). The origin of basilosaurids from protocetids has been proposed on the basis of characters basilosaurids shared with some protocetids, such as *Babiacetus* and *Georgiacetus*. Most basilosaurids have significantly more posterior thoracic and lumbar vertebrae than protocetids. Basilosaurids known from the late middle Eocene to the late Oligocene (48–36 million years ago) are found worldwide. They are distinguished as the first fully aquatic whales, as is supported by tooth isotope values (figure 2.6). This land-to-water transition was accompanied by an order of magnitude increase in body mass and oxygen-storage capacity. Diving capability, measured in terms of oxygen storage, increased from the earliest whales—estimated to range from 1.6 minutes in the wolf-sized *Pakicetus* to about 17.4 minutes in the 6,400 kg (more than 14,000 lb) *Basilosaurus isis*.

Two basilosaurid subfamilies are recognized by some workers: basilosaurines and dorudontines. Basilosaurines have elongate bodies with numerous lumbar vertebrae (20, compared with 5 in most mammals), with a maximum length of 17 m (56 ft). *Basilosaurus* was initially recognized as a reptile by the natural historian Richard Harlan, hence the name "lizard king" in his 1834 publication, but later examination by Richard Owen (1839), a distinguished professor of anatomy at the Royal College of Surgeons in England, found it to be a whale. Best known is *Basilosaurus isis*, with several hundred skeletons reported from the middle Eocene Valley of Whales in north central Egypt, approximately 170 km (105 mi) from Cairo (figure 2.14). The largest intact *B. isis* specimen (approximately 20 m [65 ft] long) is on display at the newly opened (2016) Wadi Hitan Fossils and Climate Change

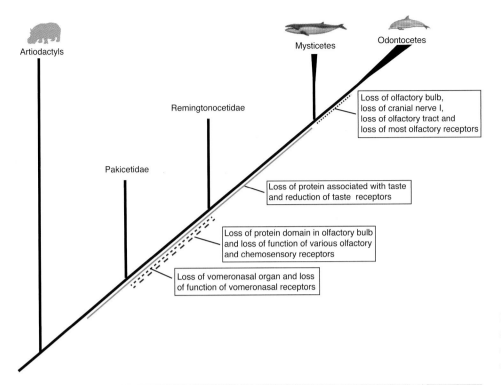

Artiodactyls

Pakicetidae

Remingtonocetidae

Mysticetes

Odontocetes

Loss of olfactory bulb,
loss of cranial nerve I,
loss of olfactory tract and
loss of most olfactory receptors

Loss of protein associated with taste
and reduction of taste receptors

Loss of protein domain in olfactory bulb
and loss of function of various olfactory
and chemosensory receptors

Loss of vomeronasal organ and loss
of function of vomeronasal receptors

Figure 2.13. Evolution of taste and smell mapped onto cetacean phylogeny. From Kishida et al., 2015.

Figure 2.14. Excavating *Basilosaurus isis* skeleton in the Valley of Whales, Egypt. Courtesy of P. Gingerich.

Locomotor Patterns, Hind-Limb Reduction, and Sexual Dimorphism

Much has been learned about the evolution of locomotion in cetaceans by the detailed anatomical study of fossils. Tail-based propulsion of extant whales evolved relatively recently from an ancestry that involved the use of both fore and hind limbs to paddle in the water. Paleontologists Hans Thewissen and E. M. Williams (2002) and Mark D. Uhen (2007) have traced the steps involved in the locomotor transition of cetaceans from land to water by mapping characters onto a phylogeny, as shown in the figure at right. The earliest whales to display pelvic paddling were ambulocetids. This was followed by reduction of the hind limb and caudal undulation, as seen in the remingtonocetid *Kutchicetus*. Protocetids employed the next stage in locomotor evolution. Skeletons reveal that most protocetids were able to move on land and in the water. Protocetids have a short lumbar region in the vertebral column, short ilium, and short femur, combined with relatively long finger and toe bones. Swimming was prob-

ably a combination of paddling with hind limbs and the tail, and they most likely spent considerable time in water hunting fast-swimming prey, much like modern sea lions. Protocetids display a broad morphological diversity in the hip region, which influenced their locomotor capabilities. For example, some protocetids, such as *Maiacetus* and *Qaisracetus*, possessed well-developed articulation between the pelvis and a solidly fused sacrum, which indicates use of the hind limbs in simultaneous pelvic paddling, as in river otters (*Lutra*). *Rodhocetus* was originally thought to have been the earliest cetacean to have a tail fluke, but other specimens and a study of the organization of the cetacean vertebral column by paleontologist Emily Buchholtz in 2007 contradict this interpretation. Differing from *Maiacetus* and *Qaisracetus*, *Rodhocetus* was found to have separate but articulating sacral vertebrae and is postulated to have been capable of simultaneous pelvic paddling (see the figure below). *Georgiacetus* and

later-diverging whales such as the protocetid *Babiacetus* and basilosaurids are hypothesized to have used dorsoventral pelvic undulation, given the loss of articulation between the pelvis and sacrum. Among currently known protocetids, *Georgiacetus* had the most flexible back, but since it lacked a tail fluke, it most likely retained the use of its hind limbs for propulsion in water, aided by undulatory movements of the posterior vertebral column.

Study of bone microstructure in stem whales by Alexandra Houssaye and colleagues in 2015 added to geologic and anatomical data on locomotor pattern. Microanatomical features such as increased bone mass, open medullary cavities, and spongious vertebrae support the contention that remingtonocetids and protocetids were shallow-water swimmers with very limited terrestrial locomotion.

The skeletal proportions of fossils can often provide information about sexual dimorphism and the mating system. The difference in body length of male and female specimens of *Maiacetus* (males 12% larger than females) has been interpreted as evidence for moderate sexual dimorphism and

Museum, a UNESCO natural World Heritage Site. *Basilosaurus cetoides*, known only from North America (and possibly Egypt), is related to the slightly smaller *B. isis*. Skeletons of *B. cetoides*, the official state fossil of Alabama, are known from rocks of the Jackson Group (Pachuta Marl and Shubuta Clay members of the Yazoo formations), formed at the bottom of a shallow Gulf Coast sea during the Eocene. A third species, *Basilosaurus drazindai*, reported from Pakistan, has been debated.

Vertebrate paleontologist Julia Fahlke and her colleagues in 2011 determined that basilosaurid and protocetid skulls are distinctly asymmetrical due to a seeming 3-D twisting of the skull (figure 2.15). A later study by Fahlke and Hampe (2015) that examined a larger sample of taxa revised this conclusion and suggested that "archaeocete" cranial asymmetry, although not conspicuous in most of the skull, may be more pronounced in the rostrum. Cranial asymmetry evolved in odontocete whales

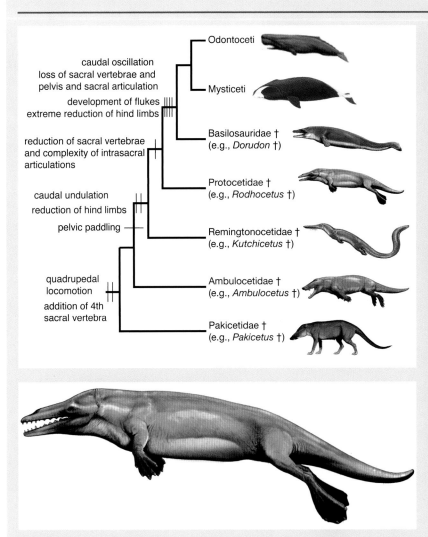

caudal oscillation
loss of sacral vertebrae and
pelvis and sacral articulation
development of flukes
extreme reduction of hind limbs

Odontoceti

Mysticeti

reduction of sacral vertebrae
and complexity of intrasacral
articulations

Basilosauridae †
(e.g., *Dorudon* †)

Protocetidae †
(e.g., *Rodhocetus* †)

caudal undulation
reduction of hind limbs

pelvic paddling

Remingtonocetidae †
(e.g., *Kutchicetus* †)

quadrupedal
locomotion
addition of 4th
sacral vertebra

Ambulocetidae †
(e.g., *Ambulocetus* †)

Pakicetidae †
(e.g., *Pakicetus* †)

(Top) Vertebral and hind-limb characters mapped onto a phylogeny of cetaceans and a reconstruction of the evolution of cetacean locomotion. Based on Thewissen and Williams, 2002; Uhen, 2007. Illustrated by C. Buell. *(Bottom)* Life restoration of the protocetid *Rodhocetus*. Illustrated by C. Buell.

probably limited male-male competition for mates. Limited opportunity to monopolize mates suggests, in turn, that food and shelter were dispersed in protocetid habitats. This is corroborated by the geographically extensive but environmentally uniform shallow marine deposits where protocetid material has been recovered.

The discovery of a "near term" fetus in a *Maiacetus* skeleton revealed positioning for head-first delivery, typical of land mammals but not whales, and the suggestion was made that early whales gave birth on land. This has been disputed and the suggestion made that the preserved smaller individual in the body cavity may instead represent a displaced fetus or a prey item in the digestive tract of the larger animal. However, the relatively large size of the "fetus" (perhaps 66 cm [2 ft] in total length, more than twice the preserved length) counters the likelihood that it was prey.

as part of a complex of traits linked to directional hearing, such as thinning of the pan bone of the lower jaws, mandibular fat pads, and isolation of the ears (figure 2.16; see further discussion of this in chapter 3).

Study of tooth microwear in basilosaurids by Fahlke and colleagues (2013) showed evidence of forceful crushing of large, hard objects such as mammalian bone. This was confirmed in a later study of bite marks in a specimen of *Basilosaurus isis* by Sniveley and colleagues (2015),

using finite element modeling. Their findings suggest that like the modern killer whale, *B. isis* applied considerable force, more than 3,600 lb at the position of an upper cheek tooth (figure 2.17). As the authors proposed, this was the part of the jaw that *Basilosaurus* used to crack skin, muscle, and bone and feed on other cetaceans, as does the killer whale, but in this case, on juveniles of the contemporaneous *Dorudon atrox*. Feeding behavior of *Basilosaurus* probably consisted of catching prey with the

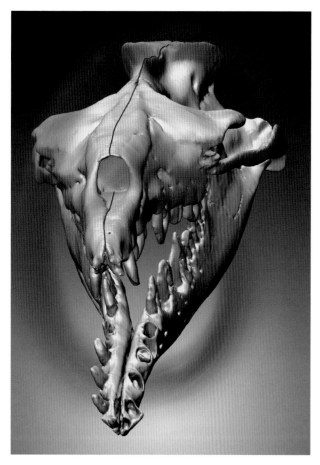

Figure 2.15. A 3-D model of *Basilosaurus isis* skull. Courtesy of J. Fahlke.

anterior teeth and breaking down prey items with the cheek teeth.

A second basilosaurid body type is seen in dorudontines, shorter and more dolphin-like. Dorudontines are the dominant whales of the late Eocene and include *Dorudon, Pontogeneus,* and/or *Cynthiacetus, Zygorhiza, Saghacetus,* and *Ancalecetus.* Although *Dorudon* is typically identified as sister taxon of Neoceti, a cladistic study by Ukrainian paleontologists Pavel Gol'din and Evgenij Zvonok (2013) positions the basilosaurid *Ocucajea* (figure 2.18)—named after the town of Ocucaje near where the specimens were found in the Eocene Ica desert of Peru—as the earliest diverging lineage within a paraphyletic Basilosauridae. This hypothesis warrants further investigation within the context of a broader taxonomic sample. One feature suggesting a closer relationship of *Ocucajea* to Neoceti is the advanced stage of telescoping (see chapter 3), with the maxilla reaching the posterior margin of the nasal. A second larger basilosaurid from Peru is *Supayacetus muizoni,* named after the Incan god of death and after Christian de Muizon of the Museum National d'Histoire Naturelle in Paris, honoring his longtime contributions to marine mammal paleontology in South America. *Supayacetus, Ocucajea,* and *Basilotritus* are

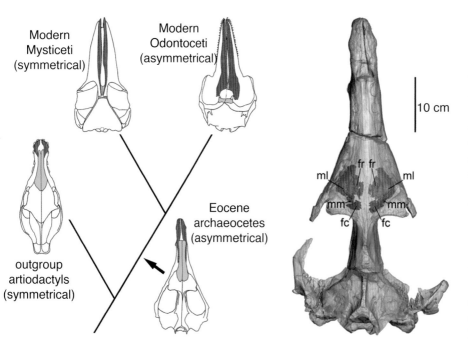

Figure 2.16. Asymmetry in the evolution of whales. *Left,* cranial asymmetry mapped onto whale phylogeny. *Right,* skull of the protocetid archaeocetes *Artiocetus clavis* (GSP-UM 3458) in dorsal view showing maxillary and frontal sinuses, visible in a 3-D micro-CT reconstruction. Note the rightward deviation of the midcranium. *Abbreviations:* fc, caudal frontal sinus; fr, rostral frontal sinus; ml, lateral maxillary sinus; mm, medial maxillary sinus. From Fahlke et al., 2011.

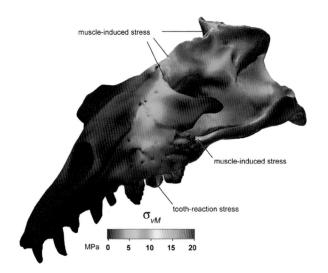

Figure 2.17. *Basilosaurus isis* skull showing bite forces (in color). *Abbreviation:* σ_{vM}, Von Mises stresses (maximum shown here of 20 MPa). From Sniveley et al., 2015.

positioned as early-diverging basilosaurids relative to more crownward *Saghacetus, Ancalecetus, Cynthiacetus, Basilosaurus,* and perhaps *Chrysocetus* and *Zygorhiza*. One of the more broadly distributed basilosaurids, *Basilotritus* is known from Africa, Europe, and North America. A distinguishing feature of this taxon is elongation of thoracic and lumbar vertebrae, which implies increased flexibility of the vertebral column as seen in extant whales.

Dorudontines are best known from North America and northern Africa, but they are also known from India-Pakistan, Europe, and New Zealand. *Zygorhiza* was originally described as a species of *Basilosaurus,* based on a braincase in the Museum für Naturkunde in Berlin. Much of our knowledge of *Zygorhiza,* known by a single species, *Z. kochii,* from the late Eocene of the Gulf Coast of North America, is based on specimens including partial skeletons described in 1936 by Remington Kellogg, longtime curator of Vertebrate Paleontology at the Smithsonian. *Zygorhiza* is smaller than *Dorudon,* with a body weight estimated at about 998 kg (2,200 lb) based on vertebral size.

Previous study suggested that *Zygorhiza* was adapted for hearing either low- or high-frequency sounds. A more recent CT scanning study of the inner ear (particularly the tight coiling of the cochlea) in *Zygorhiza,* conducted

Figure 2.18. Life restoration of two archaeocetes, *Ocucajea* (middle) and *Supayacetus* (bottom), from the Eocene of Peru. Illustrated by C. Buell.

by paleontologists Eric Ekdale and Rachel Racicot in 2015, supports the hypothesis that low-frequency sensitivity was ancestral for cetaceans and was retained in mysticetes, with subsequent high-frequency sensitivity evolving in odontocetes. This issue is not yet settled, however, since contrary to this hypothesis, adaptations for high-frequency hearing in "archaeocetes" were found in another inner ear study, by Morgan Churchill and colleagues (2016). Comparisons with additional "archaeocetes" are needed to resolve this controversy. The evolution of underwater hearing in crown cetaceans is further discussed in chapter 3.

The skeleton of the best-known dorudontine, *Dorudon atrox,* has been studied in detail by whale paleontologist Mark D. Uhen (2004). *D. atrox* was approximately 5 m (16 ft) in length with a body weight of 2,240 kg (4,938 lb), similar in size to a beluga. The eyes were directed to-

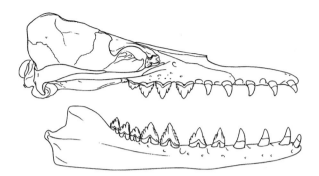

Figure 2.19. *Dorudon atrox* skull. The skull length is approximately 95 cm (3.1 ft). From Uhen, 2004.

ward the sides of the head, suggesting that these animals hunted prey underwater. The opening of the nose was positioned midway between the tip of the snout and the eyes. Cranial endocasts indicate that *D. atrox* had a tiny brain in comparison to modern cetaceans. Over the course of "archaeocete" evolution there seems to have been a doubling of brain size relative to body size over a 10 million year span, from the middle Eocene *Rodhocetus kasrani* and *Remingtonocetus ahmedi* to the late Eocene *D. atrox*, *Z. kochii*, and *Saghacetus osiris* (Gingerich, 2016). The laterally compressed elongate cheek teeth, like those of all basilosaurids, have numerous accessory denticles (cusps) and wear facets indicating that these teeth functioned in shearing prey (figure 2.19). The jaws and teeth as well as some preserved stomach contents indicate that these animals captured single prey with anterior teeth, using their posterior teeth for food processing.

Basilosaurids have forelimbs modified into flippers. Uhen (2004) described *Dorudon atrox* as having a mobile shoulder joint, like modern whales. The forelimbs were short and flattened, and the elbow joint was capable of limited bending, unlike the stiff, immobile elbow of extant whales. *D. atrox* had little mobility in the wrist, but the fingers differed from those of extant whales in having varying degrees of flexibility.

The pelvis was not attached to the vertebral column in basilosaurids, and the increased number of trunk vertebrae may have improved the efficiency of caudal undulation. In particular, the vertebral morphology of *Stromerius* from Egypt exhibits the derived condition of elongate metapophyses, attachment sites for epaxial muscles of the vertebral column. The hind limbs of basilosaurids (e.g., *Basilosaurus*, *Chrysocetus*, *Dorudon*) were tiny and were not used in locomotion. Basilosaurines such as *Dorudon* and *Cynthiacetus* had body proportions similar to dolphins and porpoises (figures 2.20, 2.21), but *Basilosaurus* had an exceptionally long body and tail, differing from the other genera in its snake-like proportions. The presence of a large, ball vertebra at the base of the tail shows that basilosaurids such as *Dorudon* had developed a fluke (a tail with triangular sides), indicating that swimming was tail-based as in extant cetaceans. Tail-based propulsion allowed cetaceans to exploit widely spaced productive areas of upwelling. In a bone microstructure study by Houssaye and colleagues (2015), basilosaurids showed specializations, similar to those of extant cetaceans, suggesting that they were more actively swimming in the open ocean.

Fragmentary, enigmatic "archaeocete" fossils have been referred to *Kekenodon onamata* (Kekenodontidae),

Figure 2.20. Life restoration of *Dorudon atrox*. From Uhen, 2004.

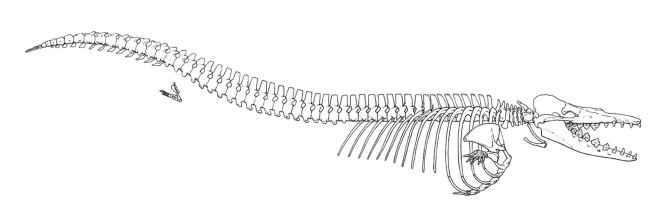

Figure 2.21. *Dorudon atrox* skeleton. Skeleton length is 4.85 m (15.9 ft). From Uhen, 2004.

from the Oligocene (28-27 million years ago) of New Zealand. Originally thought to be toothed mysticetes, they are now recognized as "archaeocetes" positioned between Basilosauridae and Neoceti.

Prorastomids and Protosirenids: Walking Sea Cows

Sirenians have a fossil record that extends from the early Eocene, 50 million years ago. The recent discovery of sirenian fossils from West and North Africa by Benoit et al. (2013) fills an important gap in our understanding of sirenians' evolutionary history. Although sirenians have been recognized as a group with African origins, the earliest record of the group up to this point was Jamaica. The sirenian earbone from Tunisia, the most primitive known, was found in a lacustrine limestone, which suggests that sirenians may have been adapted to freshwater environments before dispersing from Africa. Stem sirenians include two families: Prorastomidae and Protosirenidae (figure 2.22). Although fossil and extant sirenians are less diverse than cetaceans, the sirenian fossil record includes key transitional forms that document important macroevolutionary changes leading from terrestrial prorastomids to semiaquatic protosirenids to fully aquatic sirenians.

The oldest known stem sirenians are a paraphyletic group, the prorastomids from early and middle Eocene, 47- to 50-million-year-old rocks in Jamaica, Florida, and Africa (Tunisia). Since the oldest known sirenian locality in Chambi, Tunisia, consists of lake deposits (calcareous limestone), sirenians, like cetaceans, may have evolved first in freshwater. Sirenians later dispersed to the North Atlantic (Jamaica) from Africa, providing evidence of their subsequent adaptation to marine waters. The dense, swollen ribs of prorastomids indicate a partially aquatic lifestyle, as does their occurrence in lagoonal deposits. Richard Owen (1855) named *Prorastomus sirenoides*, meaning "forward jaw-mouth" in reference to its downturned rostrum. The skull and jaws were found in a hard gray limestone nodule in Jamaica. *Prorastomus sirenoides* was approximately 1.5 m (5 ft) in length. Unlike modern sirenians but similar to early cetaceans, prorastomids had well-developed legs and were capable of locomotion on land and in the water. Swimming was accomplished by a combination of spinal undulation and bilateral thrusts of the hind limbs. Judging from their narrow, undeflected rostrum and crown-shaped molars, prorastomids were probably browsers that fed on floating aquatic plants and, to a lesser extent, the sea grasses consumed by later-diverging sirenians. This is confirmed by the presence of sea grass macrofossils in the middle Eocene, indicating that these plants were well established in the Western Atlantic-Caribbean region. Convergent with cetaceans, prorastomids preserve evidence of isolation of the ears, indicating a specialization for underwater hearing. The nasals also exhibit retraction, a hallmark sirenian synapomorphy.

Pezosiren portelli, a pig-sized animal from the Eocene

of Jamaica with an estimated length of 2.1 m (6.9 ft), possessed a long trunk supported on four relatively short legs (figure 2.23). Similar to *Prorastomus*, the elongate, undeflected rostrum suggests it was a browser. *Pezosiren* occupied both land and water, and swam, like otters, by spinal and hind-limb undulations aided by a long tail—a convergence shared with early whales. *Pezosiren* was

found to have higher muscle oxygen-storage capabilities than living sirenians, presumably because of the higher metabolic costs entailed in functioning well both on land and in the water. However, the pachyosteosclerotic (thick, dense) ribs, which provided ballast, the retracted external nares, and the lack of paranasal sinuses show that *Pezosiren* was well adapted for aquatic life. These

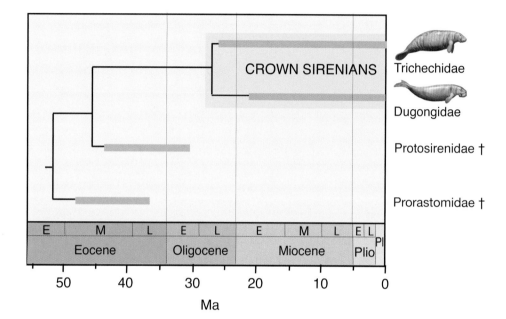

Figure 2.22. Phylogeny of stem sirenians. From Vélez-Juarbe and Domning, 2015.

Figure 2.23. Life restoration of stem sirenian *Pezosiren portelli*. Courtesy of Tim Scheirer, Calvert Marine Museum.

features, together with recovery of innominates (pelvic bones) that suggest limited terrestrial locomotion, indicate that this taxon probably spent more time in water than on land.

Protosirenids, positioned crownward of prorastomids, are known from the middle and late Eocene at localities along the Atlantic Ocean, Mediterranean Sea, and Indian Ocean. Protosirenids (*Protosiren*, *Ashokia*) are paraphyletic at the base of remaining sirenians. Protosirenids were mainly aquatic, with short forelimbs and reduced hind limbs, indicating an amphibious lifestyle. Swimming in protosirenids employed dorsoventral undulations of the enlarged tail aided by bilateral thrusts of the hind limbs. These mammals probably spent less time on land than prorastomids. Protosirenids begin to show typical sirenian cranial features such as a downturned rostrum and a broadened mandibular symphysis, indicating grazing on sea grasses near the bottom of the water column. Mea-

Evolution of Sirenian Locomotor Patterns

Like cetaceans, stem sirenians passed through a terrestrial quadrupedal stage in their transition from land to water, as shown in the figure. The earliest stage is represented by prorastomids using hind limbs rather than the tail for propulsion. The next stage, exemplified by protosirenids, is characterized by reduction of the hind limbs. The expanded caudal vertebrae of protosirenids typical of later sirenians indicate that the tail had become the major propulsive organ, although the hind limbs probably played a minor role in swimming. Crown sirenians, both manatee and, especially, dugongid lineages, show considerable variation and further reduction of pelvic and hind-limb bones; these represent the final locomotor stage: swimming with the tail only.

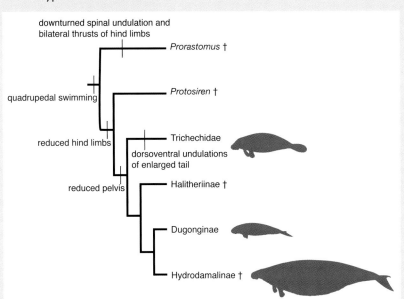

Vertebral and hind-limb characters mapped onto sirenian phylogeny and reconstruction of the evolution of sirenian locomotion. Based on Domning, 2000.

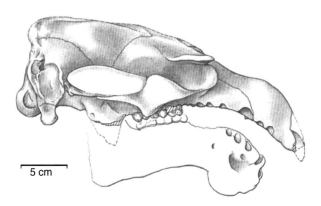

5 cm

Figure 2.24. Skull and jaw of *Protosiren fraasi*. From Gingerich et al., 1994.

surement of the isotopic composition of the teeth of protosirenids is consistent with a diet of sea grasses.

The protosirenid *Protosiren* is the best-known taxon, represented by four species. *Protosiren eothene*, from the middle Eocene of Pakistan, is the oldest and smallest protosirenid. *Protosiren fraasi* was described more than 100 years ago by the British paleontologist C. W. Andrews (1902a, 1902b, 1904), from the Mokattam Limestone of Egypt. Study of the skull of *P. fraasi* using CT scans revealed that this animal had small olfactory and optic tracts but large maxillary nerves, consistent with the diminished importance of smell and sight in an aquatic environment (figure 2.24). This, coupled with the large downturned snout, confirms the animal's enhanced tactile sensitivity. Later-diverging species include *Protosiren smithae*, from the latest middle to earliest late Eocene of the Fayum basin, Egypt, and the larger, more primitive *Protosiren sattensis*, from the late middle Eocene in Pakistan. *Protosiren* species show a variation in form and size of pelvic morphology that has been interpreted as demonstrating sexual dimorphism.

By the Oligocene, prorastomids and protosirenids were extinct. Their disappearance occurred at the same time as the origin of sirenians with an aquatic body plan belonging to the Halitheriinae and Dugonginae (see chapter 5).

3 Later-Diverging Whales

Neoceti

The next major increase in marine mammal diversity involved the crown cetaceans, or Neoceti: odontocetes, or toothed whales, and mysticetes, or baleen whales. Both lineages underwent an explosive radiation in the late Oligocene, 30–23 million years ago. Their diversification is tied to a major restructuring of the Southern Ocean ecosystem and increased production of zooplankton in areas of upwelling. The broader geologic context entailed the separation of Antarctica and South America in the late Oligocene, opening part of the Southern Ocean and allowing west-to-east flow of the newly developed Antarctic Circumpolar Current (figure 3.1, top). This current isolated Antarctica and probably allowed the Antarctic icecap to expand, global climates to cool, and oceans to become more mixed, increasing their high-latitude productivity and establishing areas of upwelling. The evolution of baleen and bulk feeding enabled mysticetes to take advantage of areas of cold, nutrient-rich upwelling. For odontocetes, high-frequency hearing, or echolocation, was a key morphological innovation that triggered their diversification. In addition, a major global sea level rise is likely to have expanded continental shelf habitats.

Among the best-known Oligocene fossil localities is the Waitaki Valley, New Zealand (33–23 million years ago), which has produced well-preserved fossils of stem mysticetes and odontocetes. (This and other fossil localities for the late Oligocene to early Miocene are shown in figure 3.1, top.) These localities span an area of tens of kilometers on the west coast of South Island, New Zealand. Faunal assemblages dominated by small to intermediate-sized mysticetes (e.g., eomysticetids) and odontocetes (e.g., squalodontids, squalodelphinids, and *Waipatia*) have been recovered from the Wharekuri Greensand and overlying younger sediments, the Kokoamu Greenstone and Otekaike Limestone. At even higher latitudes, the earliest-known mysticete, *Llanocetus*, was recovered from the late Eocene to early Oligocene (ca. 35 million years ago) La Meseta Formation in Antarctica. Other notable late Oligocene or early Miocene (24–23 million years ago) localities in the southern hemisphere include the Jan Juc Marl in coastal Victoria, Australia,

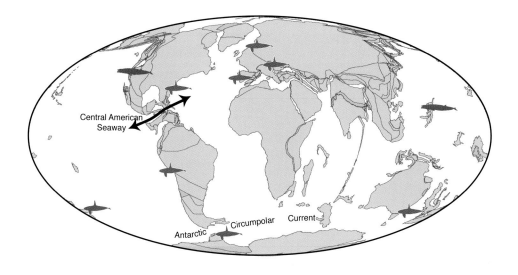

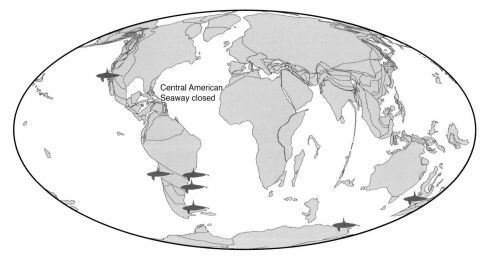

Figure 3.1. Major crown cetacean fossil localities during the late Oligocene to early Miocene (*top*) and the middle Miocene to Pliocene (*bottom*). Modified from Fordyce, 2009. Base tectonic map from www.odsn.de/odsn/about.html.

which has produced the stem toothed mysticetes *Mammalodon* and *Janjucetus*.

The North American marine Oligocene record also includes a diversity of stem odontocetes and mysticetes. Stem odontocetes (*Ashleycetus, Xenorophus, Mirocetus*) have been described from the early Oligocene of the southeastern United States—specifically, the Ashley Formation and overlying Chandler Bridge Formation near Charleston, South Carolina. This same time interval has produced other stem odontocetes such as *Simocetus* from the Alsea Formation of Oregon and stem toothed mysticetes, the aetiocetids, from the Yaquina Formation of coastal Oregon. Other aetiocetids are known from coeval deposits in Hokkaido, Japan.

Fossil localities of early Miocene age are relatively rare worldwide; they include strata of southern Argentina (Patagonia) deposited in shallow-water bays that have to date yielded only remains of cetaceans, including the oldest described right whale, *Morenocetus parvus*. Cetaceans, sirenians, and pinnipeds are found in formations of early to middle Miocene age of the Chesapeake Group of Maryland and Virginia. Early to middle Miocene localities in the Calvert Formation in the Chesapeake Group are mostly fine-grained and shallow-water sediments deposited during a subtropical climate regime. Odontocetes found here include species in at least three groups: Squalodelphinidae, Squalodontidae, and "Platanistoidea." Several species of the long-snouted eurhinodelphinids also occur, in addition to kentriodontids, delphinids, and sperm whales. Among mysticetes are "cetotheres": *Aglaocetus*, *Diorocetus*, and *Parietobalaena*.

Perhaps the greatest diversity of marine mammals occurred during the middle Miocene Climate Optimum (14–12 million years ago), which probably drove increased food production and species diversity. One of the richest marine mammal fossil localities of the middle Miocene is Sharktooth Hill in central California, where cetaceans as well as pinnipeds and desmostylians (rare) have been recovered (figure 3.1, bottom, and endpapers; see also chapters 4 and 5). Cetaceans dominate the assemblage. Among mysticetes, several "cetotheres" (*Aglaocetus*, *Parietobalaena*, *Tiphyocetus*, *Peripolocetus*) have been recovered. Odontocetes include the intermediate-sized sperm whales, *Aulophyseter*, and smaller delphinids, kentriodontids, squalodontids, and platanistids. Study of the Sharktooth Hill bone bed by Nick Pyenson and coworkers (2009), based on multiple lines of evidence (sedimentology, taphonomy), revealed that the bone bed was not formed by a catastrophic one-time event; rather, it was a long-term accumulation formed between 16 and 15 million years ago.

Neoceti Evolution
A Unique Tooth Pattern

Among mammals, whales exhibit a unique tooth pattern that is characterized by both homodonty, or simple, crowned teeth, and polydonty, an increased number of teeth. Both are associated with the loss of precise tooth occlusion related to the absence of mastication. Instead of using their teeth to chew food as do most mammals, most cetaceans use their teeth only to grab and hold prey before transport to the stomach, where breakdown and digestion occur. Brooke Armfield and colleagues' (2013) study of the evolution of tooth morphology in cetaceans and the underlying developmental controls revealed that the functional constraints underlying mammalian tooth occlusion have been relaxed in cetaceans, facilitating changes in the genetic control of early tooth development. These developmental observations are supported by the tooth morphology of fossil cetaceans. Molar complexity changed throughout the Eocene, such as the ad-

dition of cusps in stem cetaceans (ambulocetids, remingtonocetids, protocetids, and basilosaurids). Beginning in the Oligocene, stem mysticetes (e.g., aetiocetids) and odontocetes (e.g., squalodontids) underwent increases in the number of teeth (to 15) and reduced crown complexity. This differs from developments in stem whales (e.g., *Pakicetus*), which have 11 teeth and a dental formula of 3.1.4.3 in both upper and lower jaws—the same number of teeth as basal placental mammals. Extant odontocetes have a highly variable number of teeth, ranging from 0 to 50 on each side of the jaw, and no teeth erupt in adult living mysticetes. Teeth begin to develop in mysticete fetuses but are resorbed *in utero*.

Underwater Hearing

Whales lack external ears (pinnae), and they receive underwater sounds very differently than their terrestrial relatives. Whales channel sounds using specialized acoustic fats in their hollow lower jaws, and the presence of a large mandibular foramen in nearly all cetaceans suggests that this auditory pathway evolved early in cetacean evolution. Using innovative imaging methods, Smithsonian paleontologist Nick Pyenson and Maya Yamato, then a postdoc at the Smithsonian, identified an evolutionary novelty in cetaceans: an acoustic funnel that leads to the middle ear (Yamato and Pyenson, 2015) (figure 3.2). The position of the acoustic funnel appears to be significant in hearing. In some mysticetes (balaenopterids such as minke whale) the funnel faces laterally, while all toothed whales have cones that face anteriorly. The acoustic funnel is developed in crown cetaceans, suggesting it appeared at least 34 million years ago and perhaps earlier. Basal stem cetaceans (e.g., *Pakicetus*) lack specializations for underwater hearing. Stem and crown cetaceans (e.g., *Andrewsiphius*, *Remingtonocetus*, and basilosaurids) and some toothed mysticetes (e.g., *Aetiocetus*) possess anteriorly oriented tympanic bullae and acoustic funnels, suggesting that the lateral sound-reception pathway in balaenopterid mysticetes evolved more recently. One hypothesis proposes that the acoustic funnel and sound-reception pathways evolved first in stem cetaceans and were later specialized for transmitting the highly directional,

The Basic Anatomy of a Whale

Compared with the skull of a typical mammal, the skull of crown cetaceans is telescoped—a condition in which the nostrils (external nares) have moved from the tip of the rostrum to the top of the head, where they form the blowhole. Transitional stem whales document the posterior migration of the nostrils. In odontocetes, telescoping of the skull involves the premaxilla and maxilla bones (which possess upper teeth in most mammals), extending posteriorly and laterally to override the frontal bones (which form the skull roof) and crowd the parietals laterally (see the figure below). Telescoping of the mysticete skull differs from that of odontocetes in that the maxilla extends posteriorly both above and beneath the frontal bone. In the mysticete skull, the entire facial region is expanded, and the rostrum is variably arched to accommodate the baleen plates that hang from the upper jaw.

Typically, the skull of odontocetes displays cranial and facial asymmetry— a condition in which bones and soft anatomical structures on the right side are larger than those on the left. It has been hypothesized that asymmetry evolved with the right side becoming more specialized for biosonar sound production and the left side for respiration, although more recent work (Madsen et al., 2013) suggests that one side is used for production of echolocation clicks and the other side for production of whistles. Fahlke and Hampe (2015), using 3-D morphometrics in a study of fossil and extant cetaceans, confirmed that directional cranial asymmetry is an odontocete specialization, not present in mysticetes, and is probably related to echolocation (see also chapter 2). Odontocete high-frequency sounds are produced in soft tissue structures, the phonic lips, located within the nasal passages. Sound pulses, also known as clicks, are generated by air passing through the phonic lips. A large, ovoid melon is located in the facial region of odontocetes, resting on a pad of connective tissue on top of the bony rostrum of the skull. The melon contains lipids that are important in focusing sound frequencies.

The skeletons of odontocetes and mysticetes similarly show numerous adaptations for life in the water, as shown in the figure at right. The vertebral column has large spines to anchor the powerful fluke (tail) muscles, which provide propulsion. Some or all of the neck vertebrae in some whales (e.g., the bowhead, *Balaena mysticetus*) are fused, which inhibits the neck mobility important in maintaining hydrodynamic efficiency. The forelimbs are reduced and flattened into paddles, primarily employed in steering. Flipper shape varies. For example, the elongate, streamlined flippers of blue whales, *Balaenoptera musculus*, have high aspect ratios (length/width, which represents lift/drag) and function in fast swimming, whereas the short, rounded, low-aspect-ratio flippers of right whales are well adapted to their slow, continuous movement. The elbow joint of crown cetaceans is immobile; the joint between the humerus and ulna is flat instead of semicircular. Since the

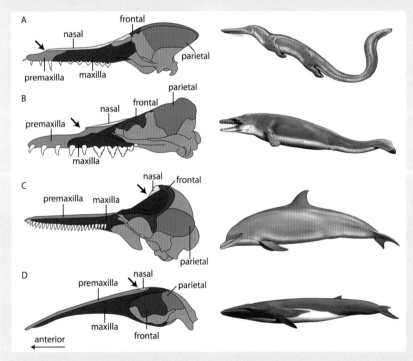

Cranial telescoping and posterior migration of bony nares in extinct and extant cetaceans. *A*, early Eocene archaeocete (Remingtonocetidae): skull of *Andrewsiphius sloani. B*, middle Eocene archaeocete (Basilosauridae): skull of *Dorudon atrox. C*, Recent odontocete: skull of *Tursiops truncatus. D*, Recent mysticete: skull of *Balaenoptera* sp. *Thick arrows* indicate locations of bony nares, and similar coloration of bones indicates homology. Skull drawings modified from *A*, Bajpai et al., 2011, fig. 11; *B*, Uhen, 2004, fig. 4; *C*, Mead and Fordyce, 2009, fig. 3; *D*, Berta et al., 2015, fig. 4.10. Life restorations of taxa illustrated by C. Buell.

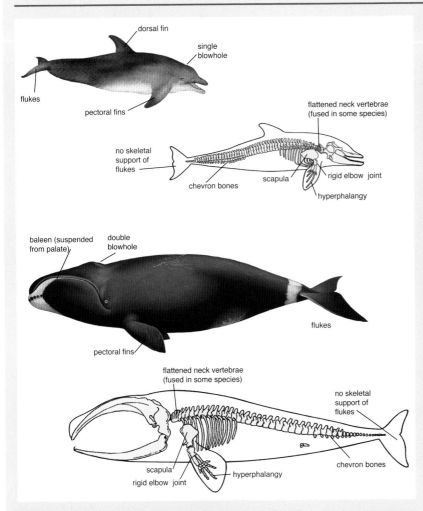

Anatomical features of an odontocete (bottlenose dolphin; *top*) and a mysticete (bowhead whale; *bottom*). Courtesy of University of Chicago Press.

elbow joint is enclosed in the flipper, the forelimb is relatively rigid. The hand exhibits hyperphalangy, in which the fingers are lengthened by additional bony elements that serve to increase the surface area of the flipper. Among mammals, hyperphalangy occurs only in cetaceans, but it is also seen in two aquatic reptile lineages (ichthyosaurs and mosasaurs). In cetaceans, hyperphalangy occurs in digits II and III in odontocetes and in digits II and IV in mysticetes. A study of hyperphalangy by Cooper and colleagues (2007) found that digit I is reduced in most cetaceans (except pilot whales, *Globicephala* spp.) and all elements of digit I have been lost in mysticetes (except balaenids). Hyperphalangy evolved 8-7 million years ago, evolving several times independently in mysticetes and odontocetes. Recent studies have expanded on anatomical studies, and the molecular basis of digit pattern control reveals that several *Hox* genes probably play an important role in digit loss and hyperphalangy.

The hind limbs of crown cetaceans are reduced to a few vestigial bones embedded in muscle. The reduction of pelvic bones and hind limbs in stem cetaceans is detailed in chapter 2. The dorsal fin is similar to the fluke in its connective tissue composition and

lack of bony support. It is uniquely developed in living mammals and varies in size and shape, serving as a keel to prevent rolling during swimming. Fossil evidence for the evolution of dorsal fins is lacking, and there are no known osteological correlates that signify its presence in extinct species.

Fins, flippers, and flukes in cetaceans, as well as in pinnipeds and sirenians, also function in thermoregulation. They contain a dense arrangement of blood vessels, countercurrent exchangers in which an artery containing warm blood from the body core is surrounded by a bundle of veins circulating cool blood from the body extremities. As the blood flows in opposite directions in these vessels, the warm arterial blood transfers most of its heat to the cool venous blood in the flipper, ensuring that there is no net heat loss. Fossil evidence for countercurrent exchangers in marine mammals is lacking, and their presence is best hypothesized based on assessments of paleoenvironmental conditions when the various lineages evolved and diversified. Also important in providing insulation is blubber, loose connective tissue that makes up the inner layer of skin. Blubber thickness varies seasonally; in bowhead whales living in polar waters it is as much as 0.5 m (1.5 ft) thick. Blubber also stores energy and functions in streamlining body contours and increasing buoyancy. As for countercurrent exchangers, the thickness of blubber in fossil cetaceans must be inferred based on indirect evidence, including the paleoenvironment, as well as the reconstructed body size of the animal.

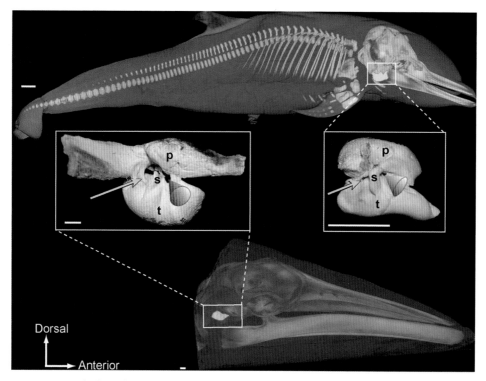

Figure 3.2. Right lateral views of CT-based 3-D reconstructions of the ears of a toothed whale, *Stenella attenuata* (*top*), and a baleen whale, *Balaenoptera acutorostrata*, juvenile specimen (*bottom*). The tympanoperiotic complex houses the middle and inner ear structures and is highlighted in *yellow*. The insert images are photographs of prepared tympanoperiotic complexes from mature individuals: *Balaenoptera bonaerensis* (*left*) and *S. attenuata* (*right*). The tympanic aperture is indicated by the *blue arrow* and the acoustic funnel is illustrated by the *pink cone*. *Abbreviations:* p, periotic; s, sigmoid process; t, tympanic. Scale bars = 2 cm. From Yamato and Pyenson, 2015.

biosonar hearing of odontocetes. As balaenopterid mysticetes modified their baleen apparatus for feeding on large aggregations of zooplankton (known as bulk feeding), their sound-reception pathways may have been displaced laterally, unconstrained by the requirements of echolocation. In addition to the presence of an acoustic funnel, extant and extinct cetaceans share a thickening (pachyostosis) of the medial wall of the auditory bulla (ear), widely assumed to be a specialization for hearing.

In addition to new anatomical insights on hearing in whales, one of the more remarkable research findings in recent years is that this ability evolved independently in echolocating bats and odontocetes from the same genetic mutations. Another example of convergent evolution is deep diving in mammals, tracked by amino acid substitutions in myoglobin, as discussed in chapter 2 (reviewed in McGowen et al., 2015).

Brain Size

Cetaceans include species with the largest brains ever to have evolved, as well as species second only to our own species in brain size relative to body size, often expressed as encephalization quotient (EQ). The evolutionary history of brain and body size in cetaceans was analyzed in a study by Montgomery and colleagues (2013) that used both extant and fossil taxa. Typically, estimates of brain size in fossils are based on measurements obtained from CT scans of the skull, although in some instances, brain endocasts—internal casts that outline the surface of the brain—are preserved. Variation in EQ has been associated with various factors, including diet, social behavior, and life history patterns. EQ values across cetaceans show opposite trends among mysticetes and odontocetes: EQ values decreased from stem to crown mysti-

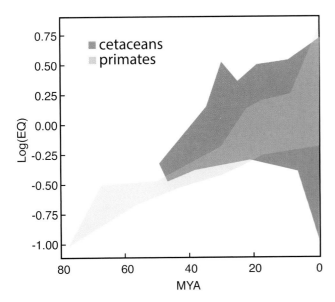

Figure 3.3. Distribution of log EQ (encephalization quotient) through time for primates and cetaceans. From Montgomery et al., 2013.

cetes, but increased from stem to crown odontocetes. The earliest odontocetes were far smaller than basilosaurid "archaeocetes," but brain size did not decrease as quickly and, as a result, EQ increased; the latter trend has been associated with the evolution of echolocation in this lineage. The decrease in EQ values for mysticetes was caused by a rapid increase in body mass, which increased at a much faster rate than brain size, resulting in the large size of today's mysticetes—all of which have relatively low EQ. Lastly, Montgomery et al.'s study points out that until only a few million years ago, the majority of mammals with the highest EQ were dolphins, not primates (figure 3.3). A later study by Gingerich (2016) cautioned against drawing conclusions about cetacean brain size without more research on brain and body size in fossil and extant whales, since there are differing methods for estimating body mass and, when body weight is underestimated, encephalization will be overestimated.

Hind-Limb Loss, Pelvic Bone Reduction, and Sexual Selection

Although crown cetaceans normally lack hind limbs, they often possess vestigial hind limbs; very rarely, an anomalous individual with hind limbs is born. The em-

bryos of crown whales, however, show the beginnings of hind limbs. Gene expression studies by Hans Thewissen and colleagues (2006) found a common developmental mechanism for the loss of hind-limb patterning in snakes and cetaceans. Results showed that hind limb bud development, a thickening of the epithelium, is initiated in cetaceans (figure 3.4), but bud growth is arrested and degeneration occurs during the fifth week of development; this is controlled by the reduction and eventual elimination of Sonic Hedgehog (*Shh*) expression, a *Hox* gene. Using the cetacean fossil record, Thewissen and colleagues were able to trace the pattern of hind-limb reduction and subsequent loss (figure 3.5). Their findings suggested that reduction of *Shh* expression may have begun 41 million years ago in basilosaurids, the earliest cetaceans to exhibit bony evidence for the tail-based propulsion characteristic of crown cetaceans. The loss of patterning of the distal hind limbs seen in basilosaurids involved losing one foot bone (metatarsal) and several toe bones. Correlated with hind-limb loss is loss of vertebral patterning, beginning with basilosaurids, which occurred after the complete loss of differentiation of lumbar, sacral, and caudal vertebrae in crown cetaceans (see also chapter 2).

In addition to the loss of hind limbs, pelvic (hip) bones also became reduced in cetaceans. The traditional explanation for reduced pelvic bones in whales is that they are vestigial structures of no use. Testing of this hypothesis by mammalogist Jim Dines and colleagues (2014) involved examining the size of pelvic bones in cetaceans in relation to mating system. They discovered that the evolution of larger pelvic bones and larger testes relative to body length seen in some whales (e.g., right whales,

Figure 3.4. Pantropical spotted dolphin (*Stenella attenuata*) embryos showing forelimb and hind-limb development. Carnegie stages (of embryonic development), *left to right:* CS 13 (earliest), CN 14, CN 16, CN 17. Courtesy of J. G. M. Thewissen.

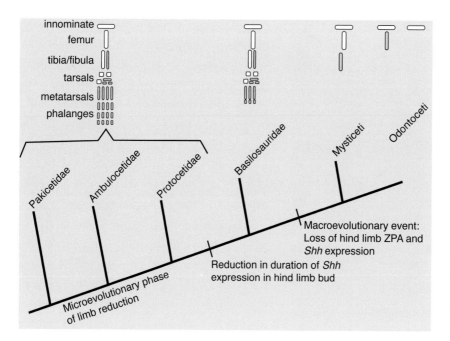

Figure 3.5. Hind-limb loss in cetaceans from the Eocene to Recent. *Abbreviation:* ZPA, Zone of Polarizing Activity. Thewissen et al., 2012.

Balaena spp., and franciscana, *Pontoporia blainvillei*) was correlated with promiscuity, a competitive mating system in which males mate with several females. Results of this study suggested that larger male genital structures (testes and pelvic bones, which anchor reproductive anatomy) were thus driven by sexual selection, with males demonstrating their fitness to females.

Body Size

Major features in the evolution of body size in Neoceti suggest that the large body size characteristic of mysticetes began relatively late in their history, during the Miocene (figure 3.6). In the late Oligocene, mysticetes (e.g., *Aetiocetus*) and odontocetes (e.g., *Simocetus*) had very similar body sizes, about 250 cm (8.5 ft) long, approximately the size of adult bottlenose dolphins (*Tursiops truncatus*). Compared with their basilosaurid ancestors (with total length estimates ranging from 485 cm [15.9 ft] for *Dorudon* to 1,600 cm [52 ft] for *Basilosaurus*), stem cetaceans near their divergence from basilosaurids were considerably smaller. By the early Miocene, body sizes for mysticetes notably increased to what seems to be their largest for that time (e.g., *Aglaocetus* at 750 cm [24.6 ft]). While odontocetes seemed to have reached

the lower limit of their size category that persists today, pronounced differences in size evolved in mysticetes between the late Miocene and the Recent. The evolution of extremely large mysticetes (>1,500 cm [49 ft] in fin and blue whales, *Balaenoptera physalus* and *Balaenoptera musculus*) is relatively recent, occurring late in the evolutionary history of cetaceans, and has been related to the evolution of engulfment feeding. This contrasts with the onset of large body size in terrestrial mammals, which occurred early in their history, 10 million years after their origin (Pyenson and Sponberg, 2011).

Diving Capabilities

In addition to the findings on diving capabilities in marine mammals that have been revealed by a molecular marker (Mirceta et al., 2013; see also chapter 1), a pathology found inside bones, called avascular osteonecrosis, has been recognized as an indicator of diving habits. This condition indicates when bone tissue has died due to lack of a blood supply. A lack of avascular necrosis indicates adaptations to prevent decompression syndrome (DCS). A study of avascular necrosis in cetaceans by paleontologists Brian Beatty and Barry Rothschild (2008) revealed that it is absent in most fossil and extant whales, espe-

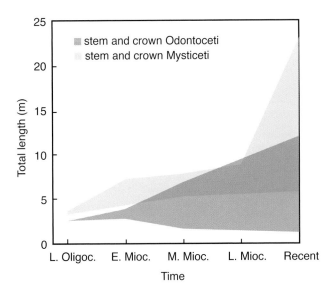

Figure 3.6. Body size evolution in cetaceans. From Pyenson and Sponberg, 2011.

cially those that are documented to be repetitive deep divers (e.g., physeterids, ziphiids). This supports the older assumptions that extant whales are physiologically adapted for avoiding DCS. The fact that basal odontocetes and crown mysticetes are exceptions and do exhibit avascular necrosis suggests that the diving physiology required for deep diving appeared early, although independently, in the two lineages. The presence of avascular necrosis in early-diverging fossil taxa in both lineages is consistent with what might be expected in taxa that were beginning to dive deeply but had not yet optimally adapted to avoid DCS.

Toothed Whales: Odontocetes
Stem Odontocetes

The phylogenetic position of a number of stem odontocetes, many included in the superfamily Platanistoidea, has been controversial, and most recent phylogenies do not find support for a monophyletic Platanistoidea, as discussed later in this chapter. Among the oldest stem odontocetes are *Cotylocara*, *Xenorophus*, *Albertocetus*, and *Archaeodelphis* from late Oligocene, 28- to 24-million-year-old rocks in North America—specifically, in the North Atlantic coastal plain (eastern United States; figure 3.1). Previously these taxa were included in the family Xeno-

rophidae. Name bearer of the Xenorophidae, *Xenorophus sloanii*, was originally described by Remington Kellogg (1923) on the basis of a partial skull from the Oligocene Ashley Formation (Rupelian) in South Carolina. Rediagnosis of this taxon by vertebrate paleontologists Al Sanders and Jonathan Geisler (2015) and Churchill et al. (2016) revealed that this family of stem odontocetes includes *Xenorophus*, *Albertocetus*, and *Echovenator*. Recent description of the inner ear anatomy of *Echovenator* from the Oligocene of North Carolina suggests that archaic odontocetes, as well as their "archaeocete" ancestors, could detect high-frequency sound (see also chapter 2). However, they probably could not hear sound in the upper range detectable by some extant odontocetes.

Cotylocara macei collected from the Oligocene (ca. 28 million years ago) Chandler Bridge Formation in South Carolina, described by Geisler and colleagues (2014), displays characters such as a dense, thick, downturned rostrum, air sac fossae, cranial asymmetry, and very broad maxillae that indicate it is one of the earliest echolocating odontocetes (figure 3.7).

Archaeodelphis, a basally positioned taxon in most phylogenies, shows a mosaic of characters resembling "archaeocetes" in its facial structures, but with broad exposure of the parietals in the intertemporal region—a key odontocete synapomorphy.

A problematic but important stem odontocete, *Agorophius pygmaeus*, based on a partial skull and tooth from

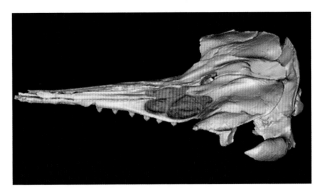

Figure 3.7. CT reconstruction of skull of *Cotylocara macei* with air sac. *Light blue* indicates excavations of an air sinus that may be homologous to the inferior vestibule as well as the route by which the sinus is connected to the soft tissue nasal passages. *Red* indicates premaxillary air sinus. From Geisler et al., 2014.

Origins of Echolocation

The evolution of odontocetes has been closely tied to the evolution of echolocation, the ability to produce high-frequency sound and receive its reflected echoes. Study of one of the best-preserved fossil odontocete ears suggests, however, that high-frequency hearing abilities developed earlier in whale evolution, about 27 million years ago, and that traits associated with this ability predate the evolution of toothed whales (Churchill et al., 2016). The oldest odontocetes show evidence in their facial structures of the ability to echolocate. Paleontologists David Lindberg and Nick Pyenson (2007) hypothesized that echolocation was initially an adaptation for feeding at night on vertically migrating cephalopods, especially nautiloids. Evidence for this coevolution between predator (whale) and prey (nautiloids) includes the finding that the gas-filled nautiloid shells produce stronger echoes than the nonshelled, soft-bodied cephalopods such as squid. These hard-shelled Oligocene nautiloids would have been easily detectable prey for early echolocating odontocetes, and this vulnerability may have been responsible for the near demise of nautiloids (as shown in the figure). Subsequent modification of echolocation as documented in the fossil record was driven by odontocetes' hunting of squid deeper in the water column.

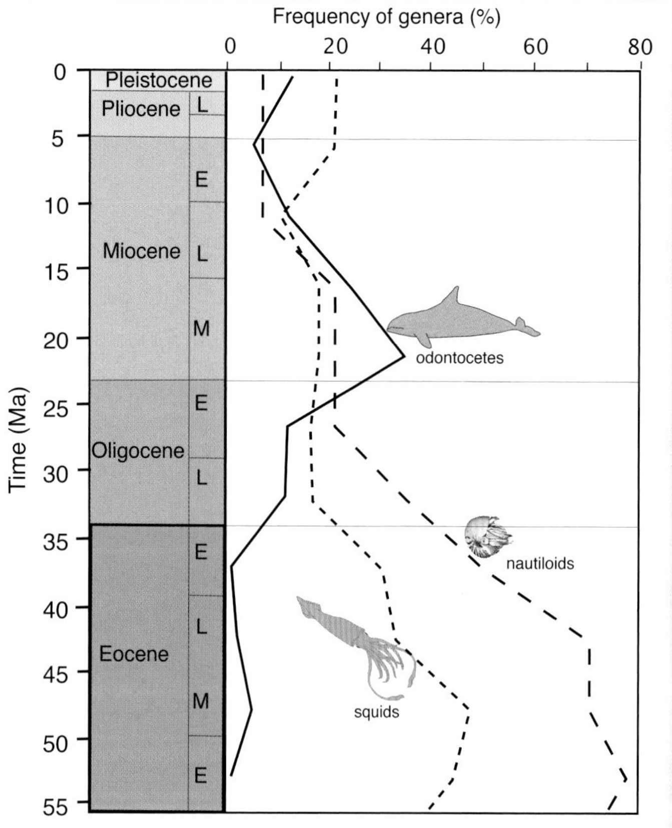

Relative frequency of predator (whale) and prey (cephalopods) over time. Modified from Lindberg and Pyenson, 2007.

the Oligocene Ashley Formation of South Carolina, has been placed in its own family, Agorophiidae. Previously this genus occupied a critical position in odontocete phylogeny because it was presumed to be the ancestral and oldest family of odontocetes. More recent discoveries of a partial skull assigned to this species (Godfrey et al., 2016), in addition to well-represented stem taxa described below, provide an improved understanding of basal odontocete morphology.

Other stem odontocetes from the early Oligocene include *Mirocetus* and *Ashleycetus*, described on the basis of partial crania, although with uncertain relationships (figure 3.8). *Ashleycetus* is another new taxon described by Sanders and Geisler (2015), from the Ashley Formation in the southeastern United States, and placed in the monospecific family Ashleycetidae. *Mirocetus riabinini*, described from probable upper Rupelian sediments of the lower Maikop beds of Azerbaijan and placed in the new family Mirocetidae, is distinguished from other stem taxa in uniquely possessing an infraorbital plate bearing alveoli, similar to that in stem cetaceans (basilosaurids and protocetids). Most importantly, these new taxa provide information on odontocete synapomorphies, particularly those involved in echolocation. Additionally, they indicate that stem odontocetes were more broadly distributed than in the eastern Atlantic.

The later-diverging, well-preserved *Simocetus rayi* (figures 3.9, 3.10), described by New Zealand paleontologist Ewan Fordyce (1994) from the early Oligocene (Rupelian) of Oregon, is placed in its own family, Simocetidae. It differs from other odontocetes in development of the nasal region, which suggests the presence of a relatively large olfactory complex. The feeding apparatus is also unlike that of other odontocetes, with downturned lower jaws that occluded against a similarly downturned, flat, edentulous anterior palate. This suggests that *Simocetus* was

a bottom-feeder preying on soft-bodied benthic invertebrates. The skulls of these stem odontocetes demonstrate that these animals had only a moderate degree of telescoping (the nares were anterior to the orbits), and the cheek teeth had multiple roots and accessory cusps on the crowns, like those of "archaeocetes" but unlike extant taxa. All taxa exhibit characters closely associated with soft tissues implicated in echolocation. One of these features is the premaxillary foramen. Based on dissections of extant odontocetes, the premaxillary foramen transmits the maxillary branch of the trigeminal nerve and is associated with nasal plug muscles, premaxillary air sac, melon, and a connective tissue theca that encloses part of the melon. Accordingly, the presence and size of the premaxillary foramen has been hypothesized to be an osteological correlate of one or more of these soft tissue structures involved in the production and transmission of high-frequency sounds (Sanders and Geisler, 2015).

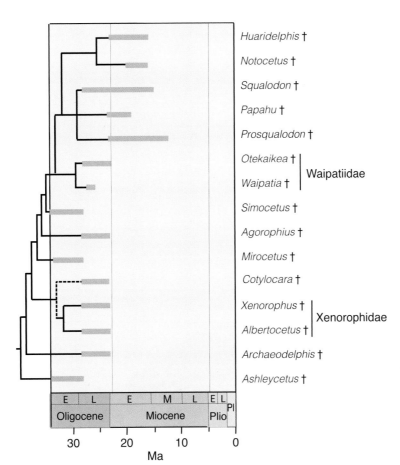

Figure 3.8. Phylogeny of stem odontocetes. Modified from Sanders and Geisler, 2015.

The squalodontids (Squalodontidae), or shark-toothed dolphins, an extinct family of odontocetes named for the presence of many triangular, denticulate cheek teeth, are known from the late Oligocene to late Miocene. They have been reported from North America, South America, Europe, Asia, New Zealand, and Australia. Squalodontids include a few species (*Squalodon*) known from well-preserved skulls, complete sets of dentition, ear bones, and mandibles, but many nominal species are based only on isolated teeth and might belong in other families. A lower jaw of *Squalodon* discovered in Malta has the distinction of being the earliest whale fossil known to science (Scilla, 1670), from the early days of paleontology (see Gingerich, 2015a). Most squalodontids were relatively large animals, 3 m (10 ft) or more in length. Their crania were almost fully telescoped, with the nares located on top of the head, between the orbits. The dentition was polydont but still heterodont, with long, pointed anterior teeth and wide, multiple-rooted cheek teeth (figure 3.9). The anterior teeth most likely functioned in dis-

play rather than in feeding, and the robust cheek teeth (with worn tips) may reflect mechanical breakdown of food items such as penguins.

The phylogenetic uncertainty of squalodontids is shared by *Prosqualodon*; it has been placed either in its own family, Prosqualodontidae, as a stem odontocete, or as sister taxon of *Waipatia* among the platanistoids. Although the affinities of squalodontids and *Prosqualodon* are debated, study of tooth morphology has suggested dietary adaptations. Moderately thick enamel is organized into features known as Hunter-Schreger bands with an outer layer of radial enamel that has been functionally related to wear resistance (figure 3.11) and may indicate a durophagous diet.

Recent discoveries of stem odontocetes, particularly from the southern hemisphere, point to a wider diversity of early-diverging taxa than previously recognized. *Huaridelphis*, named after the Huari culture (AD 500–1000), was described by Belgian paleontologist Olivier Lambert and colleagues (2014b) from the early Miocene Chilcatay Formation in the Pisco-Ica Basin of Peru (figure 3.12). *Huaridelphis* differs from other squalodontids in its smaller size, thin antorbital process of the frontal, more abrupt tapering of the rostrum, and higher tooth count. This odontocete is the smallest member of its family yet known, at less than 50 cm (1.6 ft) long. Although originally recognized as belonging in Squalodelphinidae, some recent work suggests that *Huaridelphis* should be included among the "Platanistoidea."

An increasing number of stem odontocetes have been described from New Zealand. *Awamokoa tokarahi* is from the late Oligocene (27.3–25.2 million years ago) Kokoamu

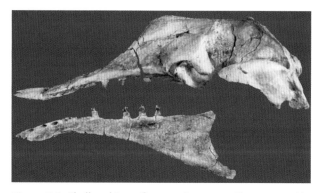

Figure 3.9. Skull and jaw of stem odontocete *Simocetus rayi* in side view. Courtesy of R. E. Fordyce.

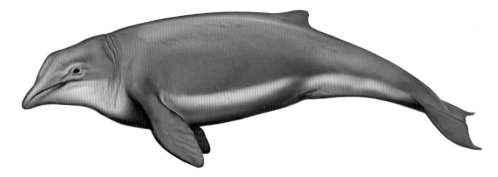

Figure 3.10. Life restoration of *Simocetus rayi*. Illustrated by C. Buell.

Greensand of New Zealand. This early-diverging taxon, which exhibits a distinct mandibular condyle, high coronoid process, shorter temporal muscle, and shorter rostrum, is interpreted to have been a raptorial feeder with a strong bite force (Tanaka and Fordyce, 2016).

Another archaic odontocete, *Waipatia*, is known by two species, *W. maerewhenua* and *W. hectori*, described by Fordyce and colleagues (Fordyce, 1994; Tanaka and Fordyce, 2015b) from the Oligocene Otekaike Limestone of New Zealand. This taxon is placed in its own family, Waipatiidae, and grouped among platanistoids, given details of the skull, ear bones, and teeth. Recent phyloge-

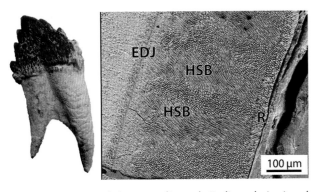

Figure 3.11. *Prosqualodon australis* tooth (in lingual view) and enamel ultrastructure. *Abbreviations:* EDJ, enamel-dentine junction; HSB, Hunter-Schreger bands; R, radial. From Loch et al., 2015.

netic analysis places *Waipatia* as sister taxon to *Awamokoa* (Tanaka and Fordyce, 2016). *W. maerewhenua* is characterized by a small (600 mm [1.9 ft]), slightly asymmetrical skull and long rostrum with small heterodont teeth (figure 3.13). The incisors are long, narrow, and procumbent. The polydont cheek teeth are more vertically inclined and have accessory denticles. *W. maerewhenua* had an estimated body size of 2.4 m (7.8 ft). *W. hectori* has a shorter skull and an estimated body size about 12% smaller than *W. maerewhenua*.

From slightly younger rocks in New Zealand (early Miocene, 21.7-18.7 million years ago, of the Kaipuke Sandstone), Gabriel Aguirre-Fernández and Ewan Fordyce (2014) have described the stem odontocete *Papahu taitapu* (Maori for "dolphin from Te tai Tapu," where it was collected) (figure 3.14). The holotype skull size indicates an animal with an estimated body length of about 2 m (6.5 ft), comparable in size to small extant delphinids (e.g., *Delphinus, Sotalia, Sousa*). Elements of the skull exhibit asymmetry, with multiple foramina on the rostrum suggestive of heavy vascularization and innervation of facial soft tissues. Pterygoid sinuses show a structurally early stage of development compared with other odontocetes; their presence is generally associated with isolation of the ears for echolocation.

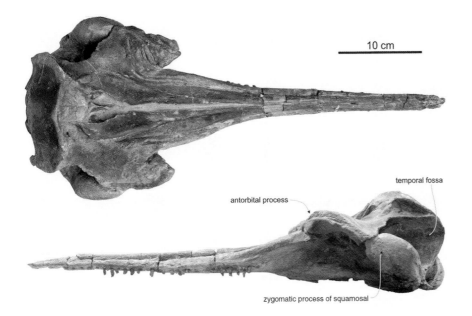

10 cm

temporal fossa

antorbital process

zygomatic process of squamosal

Figure 3.12. Top and side view of skull of stem odontocete *Huaridelphis raimondii*. From Lambert et al., 2014b.

Figure 3.13. Skull and jaw of stem odontocete *Waipatia mae-rewhenua*. Skull length is 600 mm (1.96 ft). Courtesy of R. E. Fordyce.

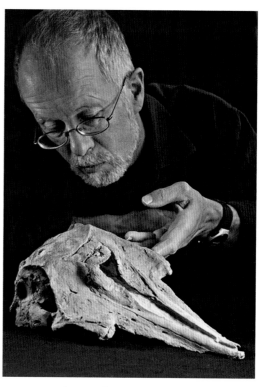

Figure 3.14. Vertebrate paleontologist Ewan Fordyce with skull of stem odontocete *Papahu taitapu*. Courtesy of R. E. Fordyce.

Recent restudy of a beautifully preserved fossil skull from late Oligocene, 23.9-million-year-old rocks (lower Miocene Kaipuke Formation) of New Zealand by Yoshihiro Tanaka and Ewan Fordyce (2014) suggests that it is not a kentriodontid, an extinct family of odontocetes (see later discussion), as previously thought. Rather, it is a new genus, *Otekaikea*, named for the source rock unit, Otekaike Limestone. *Otekaikea marplesi* was approximately 2.5 m (8.25 ft) long and had a body mass of at least 85 kg (187 lb). It had procumbent anterior teeth (tusks) and a broad, dished face for nasofacial muscles implicated in the production of echolocation sounds. The prominent condyles and unfused neck vertebrae suggest a flexible neck. A second, similarly sized species of *Otekaikea*, *O. huata*, from the Otekaike Limestone, New Zealand, was described by Tanaka and Fordyce (2015a) based on skull, ear bones, and forelimb remains. Perhaps most notable is its several long incisor tusks, the largest of which is estimated to have had a crown length of 28.0 cm (11 in). The tooth morphology and enamel ultrastructure of *Otekaikea* sp. and delphinoids are simpler and represent a transition to homodonty and polydonty, implying less demanding feeding biomechanics as seen in living odontocetes. Phylogenetic analysis suggests that Squalodelphinidae, *Waipatia*, and *Otekaikea* are either stem or crown odontocetes, with the squalodelphinids in some analyses allied with the platanistoids.

Crown Odontocetes

The phylogeny for crown odontocetes based on recent combined evidence is shown in figure 3.15. Molecular data support a radiation of the crown odontocetes in the early Oligocene, but it has been argued that for many Oligocene fossils, morphological data are not robust enough to confidently determine whether they are stem or crown odontocetes. Crown odontocetes comprise 10 families and 73 extant species and more than twice that number of described fossil taxa.

PAN-PHYSETEROIDS: PHYSETERIDAE AND KOGIIDAE

The Physeteroidea is a superfamily that today includes the families Physeteridae, the sperm whales, largest of the toothed whales, and Kogiidae, the pygmy sperm whales, and is recognized as sister group to all other living odontocetes.

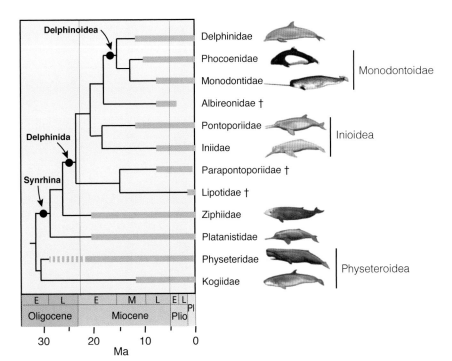

Delphinoidea

Delphinidae

Phocoenidae

Monodontidae — Monodontoidae

Albireonidae †

Delphinida

Pontoporiidae

Iniidae — Inioidea

Synrhina

Parapontoporiidae †

Lipotidae †

Ziphiidae

Platanistidae

Physeteridae — Physeteroidea

Kogiidae

| E | L | E | M | L | E | L | Pl |
| Oligocene | | Miocene | | | Plio | | |

30   20   10   0
Ma

Figure 3.15. Phylogeny of crown odontocetes. From Geisler et al., 2011.

Physeteridae: Sperm Whales. In most recent phylogenies (e.g., Boersma and Pyenson, 2015; Vélez-Juarbe et al., 2015), the earliest stem physeteroid is *Eudelphis* from the middle Miocene of Belgium, followed by other basal physeteroids (*Diaphorocetus, Acrophyseter, Idiorophus, Zygophyseter, Brygmophyseter, Livyatan, Albicetus, Placoziphius, Orycterocetus*) positioned outside the crown sperm whales and grouped in the Pan-Physeteroidea. The majority of these fossil pan-physeteroids are North American taxa, but several taxa, such as *Diaphorocetus* and *Idiorophus* from Argentina, have not been adequately studied. One recent analysis finds these two taxa nested within Physeteridae (Lambert et al., 2016) (figure 3.16). *Diaphorocetus* is characterized by small body size. *Brygmophyseter* (= *Naganocetus*) and *Zygophyseter* are the next-diverging Miocene physeteroids. *Zygophyseter* shares with all other physeteroids the characteristic deeply basined skull for housing the spermaceti organ, which suggests this is likely to be the ancestral condition. The total body length of *Zygophyseter varolai* is estimated at 6.5-7.5 m (21-24 ft), approximately 40% to 80% of the maximum length of the extant sperm whale (12.5-18.5 m [41-60 ft]). The

large, pointed upper and lower teeth of *Zygophyseter* and another physeteroid, *Acrophyseter*—in contrast to the teeth of the extant sperm whale, which are present only in the lower jaw—suggest that these stem physeteroids were active predators feeding on large prey, much like the killer whale, *Orcinus orca*. Additional evidence for macropredation comes from unusual bony outgrowths above the alveoli of the upper jaw as reported in *Acrophyseter* specimens; these outgrowths are hypothesized to have functioned as buttresses, strengthening the teeth against powerful bite forces (figure 3.17). The trend toward tooth reduction and loss in the upper jaw is seen in later-diverging physeteroids, including *Aulophyseter, Placoziphius, Scaphokogia*, and extant dwarf and sperm whales. The discovery of a stem physeteroid from Pleistocene rocks near the equator suggests that some taxa may have inhabited warm latitudes, as does the extant species.

The largest known fossil physeteroid, *Livyatan melvillei* (figure 3.18), was described by Lambert and colleagues (2010) from the 12 million-year-old Pisco Formation in Peru. It had a 3 m (10 ft) long head and 36 cm (14 in) long teeth (compared with the largest recorded teeth

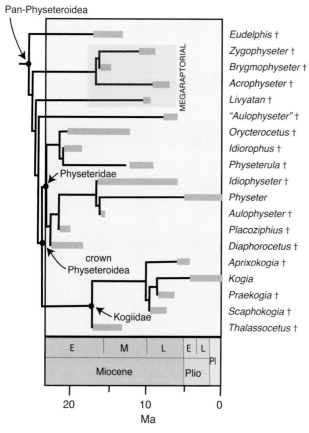

Figure 3.16. Phylogeny of Physeteridae. Modified from Lambert et al., 2016.

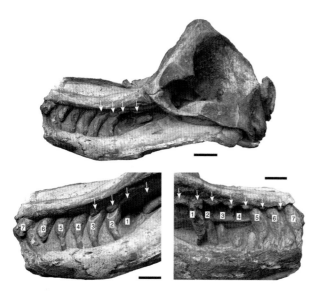

Figure 3.17. *Acrophyseter* skull and jaws showing inset of bony outgrowths above alveoli in upper jaw. Scale bar, *top* = 100 mm (3.9 in); scale bars, *bottom left and right,* = 50 mm (1.96 in). From Lambert et al., 2014a.

of *Physeter* of 25 cm [9.8 in]) and was one of the largest known predators, with an estimated body size of 13.5–17.5 m (44–55.7 ft). *Livyatan* was a top predator and probably fed on fish and on other marine mammals, including pinnipeds, odontocetes, and mysticetes. The enlarged anterior skull basin that *Livyatan* shares with the extant sperm whale implies the presence of an enlarged spermaceti organ that, in the fossil, is not associated with deep diving or suction feeding.

Albicetus oxymycterus was another large sperm whale, which together with *Brygmophyseter* and *Livyatan* is reconstructed to have had a total length of 6 m (19.6 ft) or more. *Albicetus* had lower and upper teeth; the latter, with enameled tooth caps, are unique to the genus (Boersma and Pyenson, 2015). The co-occurrence of these multiple, large, hypercarnivorous physeteroids in the middle Miocene is in sharp contrast to the single teuthophagous *Physeter* species alive today. This indicates a different structuring of marine mammal communities, with no modern analogs in today's communities, where hypercarnivory is rare.

The Physeteridae, or sperm whales, have an ancient and diverse fossil record, although only a single species, the giant sperm whale, *Physeter macrocephalus*, survives today. Sperm whales have consistently been recognized as the basal-most odontocetes. Derived characters of the skull that unite the sperm whales include, among others, a large, deep, anterior (supercranial) basin in the skull, which houses the spermaceti organ, and loss of one or both nasal bones. Various functions have been attributed to the immense sperm whale head and the spermaceti organ contained within it, ranging from its use as a battering ram to an important role in echolocation. The echolocation function of the sperm whale head is well documented based on anatomical studies by Ted Cranford and colleagues (e.g., Cranford et al., 2008a, b; 2010), but a new study based on engineering models and finite element analysis (Panagiotopoulou et al., 2016) shows that the head also evolved as a massive battering ram during male-to-male competition. Sperm whales are the largest of the living toothed whales, attaining lengths up to 19 m (62 ft) and weighing 70 tons; they are also the lon-

Figure 3.18. Life restoration of fossil sperm whale *Livyatan melvillei*. From Lambert et al., 2010.

gest- and deepest-diving marine mammal: 138 minutes and 3,000 m (>1.8 mi) on a single breath.

Uniquely found in the digestive system of sperm whales (usually associated with *P. macrocephalus* and more rarely with pygmy sperm whales) is ambergris, a substance used in making perfume during the period of commercial whaling. Previously unknown in the fossil record, fossil ambergris was reported from the early Pleistocene of Italy. Elongate, helicoidal to concentric structures containing numerous squid beaks were associated with ambergris and identified as coprolites, fossilized intestinal remains of sperm whales (Baldanza et al., 2013).

The fossil record of the Physeteridae goes back at least to the Miocene (late early Miocene, 21.5–16.3 million years ago), and earlier if the fragmentary *Fereceto-therium* from the late Oligocene (23+ million years ago) of Azerbaijan is included. By middle Miocene time, physeterids were moderately diverse. The family is fairly well documented from fossils found in South America, North America, western Europe, the Mediterranean region, Australia, and New Zealand.

Kogiidae: Pygmy and Dwarf Sperm Whales. The family Kogiidae, one of the rarest lineages of toothed whales, includes the pygmy sperm whale, *Kogia breviceps*, and the dwarf sperm whale, *Kogia sima*. They are closely related to the sperm whale family (Physeteridae) and have a worldwide distribution. Pygmy sperm whales are appropriately named because males attain a length of only 4 m (13 ft), and females are no more than 3 m (10 ft) long—at most, one-fifth the size of their physeterid cousins. The dwarf pygmy sperm whale is even smaller, with adults ranging from 2.1 to 2.7 m (7-9 ft). As in physeterids, the skull exhibits a large anterior basin for the spermaceti organ, but kogiids differ markedly in their small size, short rostrum, and other details of the skull. The smaller spermaceti organ of kogiids suggests less intense sexual selection, which is supported by their reduced pelvic bones in comparison with the extant sperm whale.

The oldest kogiids are from the early Miocene (8.8-5.2 million years ago) of Belgium. Five fossil genera have been described: *Thalassocetus* from the early middle Miocene of Belgium, *Scaphokogia* from the late Miocene of Peru, *Praekogia* from the late Miocene of Baja California, Mexico, *Aprixokogia* from the early Pliocene of North Carolina, and *Nanokogia* from the late Miocene of Panama. The extant genus *Kogia* has a fossil record that extends back to the late Pliocene of Italy. Early-diverging kogiids include *Thalassocetus*, *Aprixokogia*, and *Scaphokogia*. *Nanokogia* is a later-diverging taxon and, along with *Praekogia*, is more closely related to the extant genus *Kogia* than to other taxa (Lambert et al., 2016; Vélez-Juarbe et al., 2015) (figure 3.16). Based on morphological features shared by *Nanokogia* and *Kogia* (i.e., short rostrum), as well as paleontological and geologic evidence (e.g., abundance of fish and estimates of water depth), it is hypothesized that the fossil sperm whale suction-fed on fish and squid. The spermaceti organ was probably larger than in *Kogia*, suggesting that it may have had different sound-generating capabilities than in the extant pygmy sperm whale (Vélez-Juarbe et al., 2015).

SYNRHINA: ZIPHIIDAE, PLATANISTOIDEA, AND DELPHINIDA

A clade of crown odontocetes, the Synrhina (derived from the Greek for "together nose," referring to the merged soft tissue nasal passages distal to the external bony na-

res), was proposed by Geisler and colleagues (2011) to include the most recent common ancestor of Ziphiidae, Platanistidae, and Delphinida.

Ziphiidae: Beaked Whales. Ziphiidae are second only to delphinids in species diversity, with six genera and 22 extant species, although they are among the world's least-known large mammals given their elusive, deep-diving habits (up to 3,000 m). They are characterized by a long, slender snout that is frequently drawn out into a beak and from which the group obtains its common name, beaked whales. Although an alliance of Physeteroidea with Ziphiidae is recovered in some morphological analyses, most recent analyses do not support this arrangement. Other distinguishing features of extant beaked whales include pronounced sexual dimorphism, elevation of the narial region, one pair of anteriorly converging throat grooves, and extreme tooth reduction. One evolutionary trend in ziphiids is toward the loss of all teeth in the rostrum and most in the mandible, with the exception of one or two pairs of teeth at the anterior end of the jaw that erupt and become much enlarged, together with other skull specializations only in males.

Other strange structures possessed by both extant and fossil (see below) ziphiids are various bony structures of the facial region, rostrum, and mandible, many of them located deep in soft tissues. These structures have been hypothesized to serve as weapons, sound transmitters, and/or ballast during deep dives, although for extant taxa they are usually explained as resulting from sexual selection—serving as a means of visual display—which is likely, since they are typically absent in females. Paleontologist Pavel Gol'din (2014) presented an intriguing hypothesis that some of these bizarre facial structures may function in "auditory" display. Since the density difference of these bony structures provides a reflective difference that would be detected by echolocation, they may play a role in social interactions and in individual or species recognition. As noted by Gol'din, this "echoic imaging" hypothesis has been used to explain the co-occurrence of deep-diving ziphiid species, such as *Mesoplodon* spp., differing only in invisible or barely visible structures.

Ziphiids inhabit offshore waters of deep ocean ba-

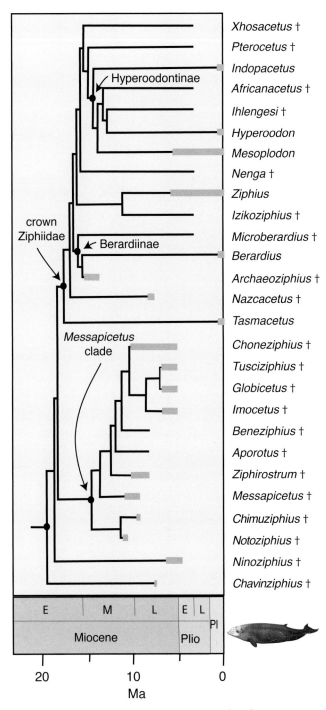

Figure 3.19. Phylogeny of Ziphiidae. Modified from Bianucci et al., 2016.

in a higher number of known fossil species than of extant species (reviewed in Bianucci et al., 2016). Ziphiids first appeared in the North Sea, with a more controversial specimen dating from the early Miocene. The archaic ziphiid *Archaeoziphius microglenoideus* from Antwerp has been used to calibrate the divergence event in crown ziphiids, dated in the middle Miocene (15-13.2 million years ago) (figure 3.19). This small ziphiid, described on the basis of several partial skulls, was much smaller than *Berardius* spp. and probably not larger than 4-5 m (13-16 ft) in length. Phylogenetic analysis by Italian paleontologist Giovanni Bianucci and colleagues (2016) divides ziphiids into two clades: the *Messapicetus* clade, which together with other stem ziphiids is known from the southeastern Pacific and the North Atlantic, and crown Ziphiidae, predominately from the Southern Ocean (figure 3.19). These scientists noted that the two lineages share similar evolutionary trends, including tooth reduction, increased thickness of rostral bones, changes in facial morphology, and increased size.

In an unusual fossil occurrence, more than 40 ziphiid skulls were recovered from dredgings of the Atlantic sea floor off the coasts of Portugal and Spain. Study of these specimens by Bianucci and colleagues (2013b) led to the recognition of several new genera and species. *Globicetus hiberus* is an especially strange-looking ziphiid that has a large spherical ossification in the middle of its rostrum, formed by fused premaxillae (figure 3.20). Another bizarre new ziphiid, *Imocetus piscatus*, bears an oddly

Figure 3.20. Skull of fossil ziphiid *Globicetus hiberus* illustrating the spherical premaxillary prominence. Rostrum length is 530 mm (1.7 ft). From Bianucci et al., 2013b.

sins, and much of our knowledge of them comes from strandings. Previously they were poorly known in the fossil record, but discoveries in the past few decades from the middle Miocene and Pliocene of Europe, North and South America, Africa, Japan, and Australia have resulted elongate facial region and elongate maxillary ridges, interpreted as areas of origin for facial and rostral muscle acting on the nasal passages, blowhole, and melon. The geologic age of these dredged specimens is uncertain, but they are most likely from late early to middle Miocene sediments (figure 3.19). Also recovered from dredgings was a ziphiid assemblage off the South African coast that resulted in description of 10 new species in nine genera (Bianucci et al., 2007). Comparison of these dredging assemblages reveals substantial faunal differences. These cetacean compositional differences may reflect the presence of equatorial warm water that acted as a barrier to dispersal between fossil ziphiids off the Iberian Peninsula and South African coasts.

Mesoplodon, with 15 extant species, is the most speciose of any known cetacean genus (see also Berta, 2015). By contrast, only two fossil species of *Mesoplodon* are recognized: *M. slangkopi*, based on skulls dredged from the ocean floor off South Africa, and *M. posti* from Belgium. Specimens of *M. posti* have been dated to between 4.86 and 3.9 million years ago, and because phylogenetic analysis nests *M. posti* among extant species of the genus, the timing of the radiation of the extant species is constrained to very early in the Pliocene (Zanclean).

The South American fossil record of ziphiids is diverse and includes six genera. Five species are known from the Miocene-Pliocene of Peru, suggesting that the nutrient-rich coastal waters of the southeastern Pacific were an important area of radiation of stem ziphiids. One of the best-known fossil ziphiids, *Ninoziphius platyrostris* (figure 3.21), from the early Pliocene, was less specialized for suction feeding than most extant ziphiids, and tooth wear in the holotype indicates benthic feeding. Suction feeding, however, was acquired early in the evolutionary history of ziphiids. Another bizarre ziphiid from Peru, *Nazcacetus urbinai*, possessed a long rostrum and a pair of large anterior teeth in the lower jaw, similar to crown ziphiids, suggesting that sexual dimorphism was present early in ziphiid evolutionary history.

The third Peruvian fossil ziphiid is the long-snouted *Messapicetus* from the middle Miocene, known also from the eastern Atlantic (United States and Italy) (figures 3.22,

3.23, 3.24). Discovery in Peru of a skeleton of *Messapicetus gregarius*, estimated at 4.1–4.5 m (13.4–14.7 ft) in length and weighing 1,842 kg (4,000 lb), together with a bony fish assemblage, provides evidence of a predator-prey relationship and offers important clues to the ancestral habitat of ziphiids. Researchers believe that the whale captured and consumed a large number of sardine-like fish, then a short time later died (perhaps due to ingesting a toxin), sank to the ocean floor, and was entombed, but not before disgorging some of the contents of its last meal (figure 3.24). Another long-snouted beaked whale, *Dagonodum mojnum*, recently described from the upper Miocene (9.9–7.2 million years ago) Gram Formation of Denmark, was identified in a phylogenetic analysis as sister taxon of *Messapicetus*. Findings of *Messapicetus*, *Ninoziphius*, and *Dagonodum* suggest that stem ziphiids were less specialized for suction feeding and deep diving than are extant ziphiids and were more likely living at shallower water depths and feeding on fish. One of the basalmost ziphiids, and the first record of ziphiids from the southwest Atlantic, is *Notoziphius bruneti* from the late Miocene Puerto Madryn Formation of Patagonia, Argentina. Among the more diagnostic elements of the partial skull and only known specimen of this species are the large, triangular, and markedly asymmetric nasals.

Two new ziphiid genera and species, *Chavinziphius maxillocristatus* and *Chimuziphius coloradensis*, are from the Pisco Formation exposed at the well-dated marine vertebrate localities of Cerro Colorado and Cerro Los Quesos. The Tortonian-aged *Chimuziphius* is nested in the *Messapicetus* clade (figure 3.19), while the Messinian-aged *Chavinziphius* is more basally positioned relative to this clade and recognized as the earliest-diverging ziphiid (Bianucci et al., 2016).

Another unusual occurrence of a large ziphiid (estimated at 7 m [23 ft] in length) is a skull from early to middle Miocene, 17-million-year-old freshwater sediments in Kenya. As is true for extant whales, ziphiids occasionally enter rivers. The size of the ziphiid, geologic context of the locality, and associated fauna indicate that it may have stranded in the river while migrating along the east coast of Africa.

Figure 3.21. Life restoration of fossil ziphiid *Ninoziphius platyrostris*. From Lambert et al., 2013. Photo courtesy of C. Letenneur.

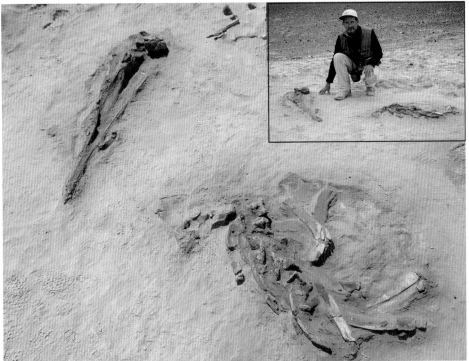

Figure 3.22. Partial skeleton of fossil ziphiid *Messapicetus* at Cerro Colorado, Peru. Courtesy of G. Bianucci.

Figure 3.23. Life restoration of *Messapicetus gregarius* preying upon a school of aged sardines. From Lambert et al., 2015a. Illustrated by A. Gennari.

Platanistoidea: Allodelphinidae, Squalodelphinidae, and Platanistidae (Ganges River Dolphin). The extant Asian River dolphin family, Platanistidae, and the extinct Squalodelphinidae and Squalodontidae were originally included in the Platanistoidea. Later, other extinct taxa (e.g., *Prosqualodon, Dalpiazina, Allodelphis, Eurhinodelphis, Zarhinocetus*) and extant river dolphin lineages (iniids, lipotids, pontoporiids) were included in this superfamily. More recently, the concept of platanistoids has been expanded to include southern hemisphere extinct taxa: *Huaridelphis, Otekaikea, Papahu,* and the Waipatiidae (discussed earlier as stem odontocetes). More systematic work, currently underway, is clearly needed to resolve relationships among "platanistoids." The concept of Platanistoidea as including three groups—Allodelphinidae, Squalodelphinidae, and Platanistidae—is the current consensus view, with

inclusion of the fossil Squalodontidae and Waipatiidae in this clade more controversial. I have followed Boersma and Pyenson's (2016) concept of Platanistoidea since it is the most comprehensive to date, including all putative platanistoid lineages. The extant Asiatic river dolphin, *Platanista gangetica* (the blind endangered Ganges and Indus river dolphins recognized as distinct subspecies), constitutes the family Platanistidae. Members of this taxon are characterized by a long, narrow rostrum, expanded frontal ridges, numerous narrow, pointed teeth, and broad, paddle-like flippers.

Fossil records of this family were virtually unknown until the first record of a fossil platanistine (the subfamily that includes the extant *Platanista*), based on a diagnostic element (a petrosal), was reported by Bianucci and colleagues (2013a) from the Miocene sediments in the Ama-

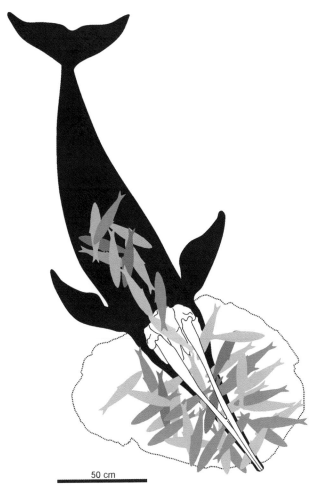

Figure 3.24. *Messapicetus gregarius* and fish associated with skeleton. From Lambert et al., 2015a.

50 cm

(Platanistinae), although they differ in rostral profiles and cranial symmetry and in their development of pneumatized bony facial crests.

Several early Miocene odontocete genera such as *Notocetus*, with small, slightly asymmetrical skulls, moderately long, tapered rostra, and near-homodont single-rooted teeth, are typically either included in the Squalodelphinidae or placed as adjacent stem odontocetes, including the platanistoids (Geisler et al., 2011). Although squalodelphinids are relatively rare in the fossil record, they were broadly distributed in the North and South Atlantic and the South Pacific.

The family Allodelphinidae was previously known by two species, *Allodelphis pratti* and *Allodelphis woodburnei*, named by Larry Barnes, both from the early to middle Miocene of California. Later a second taxon, *Zarhinocetus*, was assigned to this family, and most recently, two new early Miocene genera, *Goedertius* from Oregon and *Ninjadelphis* from Japan, have been allocated to this family. The rostra of allodelphinids are extremely elongate and flattened dorsoventrally, with a sulcus along the premaxilla-maxilla suture; there are numerous small teeth and a mandible that extends to the anterior edge of the rostrum. Members of this taxon were relatively small, less than 4 m (13 ft) in total body length. The neck vertebrae are highly elongated anteroposteriorly (unlike in modern odontocetes), which, combined with a large occipital shield and thick nuchal crest, suggests that allodelphinids had well-developed neck muscles and were capable of considerable movement of the head. Enhanced head movements have been hypothesized as a raptorial feeding adaptation similar to the *Platanista* style of head bobbing while capturing prey. Several allodelphinids are included in a recent phylogenetic analysis (Boersma and Pyenson, 2016), which positions *Allodelphis* and *Arktocara* as sister taxa most closely related to *Zarhinocetus* and *Goedertius*.

DELPHINIDA: KENTRIODONTIDAE, LIPOTIDAE (BAJI), AND INIOIDEA

Delphinida was originally proposed to include most river dolphins (iniids, lipotids, platanistids) as well as physe-

zon Basin of Peru. This discovery showed that platanistines were not confined to Asia. Given the prior tentative assignment to this subfamily of a fragmentary jaw from Oregon, Bianucci and colleagues suggest that the early phase of platanistine evolution was not restricted to fresh water. Platanistines are hypothesized to have inhabited the North Pacific, where they had a marine, North Pacific origin before or during the early Miocene. Some platanistines then entered the freshwater Amazon Basin before the end of the middle Miocene. Eventually, platanistines reached the Indian Ocean and the Ganges and Indus Rivers, where they survive at present.

Middle to late Miocene, 16- to 6-million-year-old marine species of *Zarhachis* and *Pomatodelphis* (grouped in the Pomatodelphininae) are closely related to *Platanista*

terids, ziphiids, and delphinoids (Muizon, 1984). Most recent phylogenies, however, support a more inclusive clade composed of fossil kentriodontids, lipotiids, inioids, pontoporiids, and delphinoids (e.g., Pyenson et al., 2015) (figure 3.25). Among stem "dolphins" of the Miocene is the extinct family Kentriodontidae. Kentriodontids were small to moderately large animals, with numerous teeth, elaborate basicranial sinuses, and symmetrical cranial vertices (figure 3.26). Kentriodontids were diverse, with 17 described genera recorded from Miocene sediments. *Kentriodon pernix*, the type specimen of *Kentriodon* based on several skulls from the Calvert Formation in Maryland, was originally assigned to the family Delphinidae but later recognized by Larry Barnes (1978), longtime vertebrate paleontologist at the Natural History Museum of Los Angeles County, as belonging in the family Kentriodontidae. This taxon was recovered from middle Miocene sediments dated at 19.5–18.5 million years old and has been used as a calibration point for the Delphinida. *Kentriodon* is variously positioned outside Inioidea + Delphinoidea or as an early stem delphinoid. Other studies recognize "kentriodontids" as paraphyletic, a taxonomically diverse, grade-level family of geographically widespread oceanic dolphins consisting of four or five genera. The most recent analysis that included the largest number of "kentriodontids" to date (Lambert et al., 2017) confirmed paraphyly of the group, but it found that "kentriodontids" were excluded from Delphinoidea and instead were distributed among five different extinct lineages and/or clades that are positioned as successive sister groups to Inioidea + Delphinoidea. Kentriodontid diversity declined later in the Miocene, and the last kentriodontids were contemporaneous with early delphinids, their likely ecological replacements.

The baiji, or Chinese river dolphin, *Lipotes vexillifer*, inhabited the Yangtze River, China. It possessed a long, narrow, upturned beak, a low, triangular dorsal fin, broad, rounded flippers, and very small eyes. The last documented sighting of this species was in 2002, and it has now been declared extinct (IUCN Red List of Threatened Species 2014, www.iucnredlist.org), a victim of massive habitat degradation.

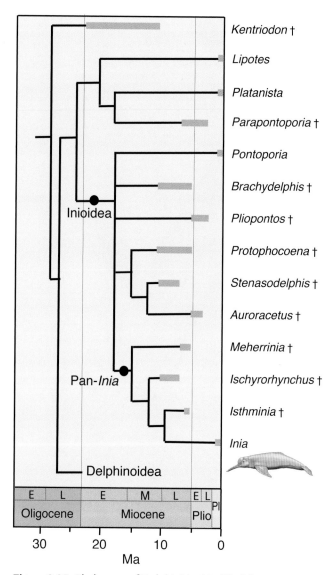

Figure 3.25. Phylogeny of Delphinida. Modified from Pyenson et al., 2015.

The purported fossil lipotid *Prolipotes*, based on a fragment of mandible from China, is too incomplete to be confirmed as belonging to this taxon. Although originally thought to be similar to the La Plata river dolphin and placed in the Pontoporiidae, most cladistic analyses (e.g., Geisler et al., 2011) identify the sister taxon of *Lipotes* as the long-snouted dolphin *Parapontoporia*, reported from marine and nonmarine sediments of late Miocene to middle Pleistocene 8.6- to 6-million-year-old deposits in the North Pacific (California, Baja California) (figure

3.27). In some phylogenies (e.g., Pyenson et al., 2015), *Parapontoporia* is grouped in a larger clade with *Platanista* and *Lipotes* (figure 3.25). *Parapontoporia* is known by three species, the oldest of which is *P. pacifica* from Baja California. *Parapontoporia* is characterized by a skull with an extremely long, slender rostrum and a lower jaw with numerous small, sharply pointed teeth (80–82 per side in each jaw). A possible occurrence of *Parapontoporia* in lower Pliocene sediments in Japan indicates a late Miocene or early Pliocene dispersal to the western North Pacific.

Given that *Lipotes* was an occupant of fresh water, the question remains as to when and where parapontoporiids became adapted to fresh water. According to Boesse-necker and Poust (2015), discovery of a petrosal of *Parapontoporia* sp. in the nonmarine late Pliocene to early Pleistocene Tulare Formation of the San Joaquin Valley, California, hints at a past freshwater life for this lineage. The geologic history of this region indicates that at the time of death, this individual occupied a lake or brackish water fed by rivers.

Inioidea: Iniidae (Bouto) and Pontoporiidae (Franciscana). Inioidea is defined as the clade consisting of the last common ancestor of the river dolphins *Inia* and *Pontoporia* and all of its descendants (Geisler et al., 2012; Pyenson et al., 2015) (figure 3.25). The pink dolphin or bouto, *Inia geoffrensis*, is a freshwater species with reduced eyes that

Figure 3.26. Articulated skeleton of kentriodontid-like dolphin at Cerro Colorado, Peru. Courtesy of G. Bianucci.

Figure 3.27. Skull and jaws of *Parapontoporia sternbergi* at San Diego Natural History Museum. Courtesy of T. Deméré.

is found only in the Amazon and Orinoco river drainages of Brazil, Peru, and Ecuador. The name "bouto" comes from the sound of its blow. The unique pink coloration may be due to the temperature and iron content of the water. A second extant species based on molecular data, *Inia boliviensis*, is recognized from the Bolivian Amazon Basin. And a third species, *Inia araguaiaensis*, from the Araguaia River basin of Brazil, has been described based on both morphological and molecular data (Hrbek et al., 2014), although questions remain about very limited sample sizes. Iniids (including fossil taxa) are characterized by an extremely elongated rostrum and mandible, a very narrow supraoccipital, a greatly reduced orbital region, and pneumatized maxillae forming a crest. Dentally, iniids are identified by conical front teeth and molariform posterior teeth, the latter feature an adaptation for crushing armored fish.

Like pontoporiids, iniids were more diverse in the past. The fossil record of iniids goes back to the late Miocene of South America and the early Pliocene of North America. *Meherrinia isoni*, named for the Meherrin River where all known specimens were found in marine rocks of the late Miocene of North Carolina, is tentatively referred to the Iniidae. Previous studies suggested that iniids originated in South America, but given the North American record of *Meherrinia*, a North Atlantic origin for the group is also a possibility. Cladistic analysis has positioned *Meherrinia* in the clade Inioidea within Argentinian taxa, including two genera from the late Miocene Ituzaingo Formation: *Ischyrorhynchus* and *Saurocetes* (Geisler et al., 2012) (figure 3.25). From the Brazilian Pleistocene, a single undescribed new species is likely to be a valid *Inia*, sp. nov., from the Rio Madeira Formation (Cozzuol, 2010). Of these extinct taxa, only *Ischyrorhynchus vanbenedeni* is known by multiple, relatively well-preserved skulls and rostral fragments with teeth.

A new genus and species of inioid, *Isthminia panamensis*, was described by Pyenson and colleagues (2015) from the Pina Facies of the late Miocene of the Chagres Formation, Panama (figure 3.28). *Isthminia* was collected from marine rocks deposited before the Isthmus of Pan-

ama formed approximately 6 million years ago. *Isthminia* is one of the largest known inioids, with a total length calculated between 284 and 287 cm (9.3–9.4 ft). Its size is similar to that of medium-sized to large extant delphinoids such as Risso's dolphin, *Grampus griseus*, which has a total length averaging 283 cm (9.2 ft). Sedimentological and taphonomic data indicate that *Isthminia* was a marine species, and analysis of tooth morphology suggests that *Isthminia* shared a similar ecology with modern oceanic delphinoids. Phylogenetic analysis positions *Isthminia* as the sister taxon of *Inia* in a larger clade that includes *Ischyrorhynchus* and *Meherrinia* (figure 3.25) and is consistent with a broader marine ancestor for inioids. Further comprehensive phylogenetic analysis is warranted, since an alternative hypothesis positions a recently described new inioid from the Miocene Pisco Formation of Peru (Lambert et al., 2017) as sister to *Inia*.

The only definitive taxon from the United States that is widely considered to be an inioid is the enigmatic *Goniodelphis hudsoni*, represented by a poorly preserved cranium from the late Miocene Bone Valley Formation of Florida. This fossil locality is probably marine, but both terrestrial and marine mammals have been recovered from the Bone Valley Formation, so it is possible that *Goniodelphis* inhabited a marine or fluvial environment. Other fossil iniids (e.g., *Ischyrorhynchus*, *Saurocetes*) have been reported from fluvial deposits.

The Franciscana, or La Plata dolphin, *Pontoporia blainvillei*, which lives in estuarine and coastal waters in the western South Atlantic of Brazil, Uruguay, and Argentina, is the only living species of the family Pontoporiidae. Both morphological and molecular data support pontoporiids as most closely related to iniids. With a total body length ranging between 1.2 and 1.7 m (3–5 ft), *P. blainvillei* is one of the smallest living dolphins. Pontoporiids have a more diverse fossil record with at least six described species from both freshwater and marine deposits. They seem to have first evolved during the Miocene, and fossil species are described from late Miocene to early Pliocene (6–3 million years ago) marine deposits in South America, the eastern coast of North America, and the North Sea.

Figure 3.28. *Top*, lateral view of type skull of *Isthminia panamensis*; skull length is 571+ mm (1.87+ ft). A 3-D model of the skull is on the Smithsonian Institution website (http://3d.si.edu). *Bottom*, life restoration of *Isthminia*. From Pyenson et al., 2015.

Most pontoporiids have long, narrow rostra and multiple tiny teeth. Although Pliocene taxa have virtually symmetrical cranial vertices, earlier pontoporiids exhibit asymmetrical skulls with projection of the maxilla onto the lateral side of the nasal.

Brachydelphis, placed in the monogeneric subfamily Brachydelphininae, is known from several species in Peru and Chile and characterized by both long- and short-snouted forms. The latter is more specialized for suction feeding, whereas the long-snouted form with a higher tooth count was probably more adapted for raptorial feeding (Geisler et al. 2012; Lambert and Muizon, 2013). Other extinct South American pontoporiids in-

clude *Pontistes* from the late Miocene Parana Formation of Argentina and the Bahia Inglesa Formation in Chile and *Pliopontos* from the early Pliocene part of the Pisco Formation in Peru.

The fossil pontoporiid *Stenasodelphis russellae* was described by Godfrey and Barnes (2008) on the basis of a partial cranium from the late Miocene (ca. 10-9 million years ago) St. Mary's Formation of Maryland. This species is characterized by its asymmetrical cranial vertex, elevated nasal, and transversely expanded premaxillary fossae. Another fossil pontoporiid, *Auroracetus bakerae*, reported by Gibson and Geisler (2009), is based on an incomplete skull from the Yorktown Formation of the early

Pliocene (4.0-3.8 million years ago) of North Carolina. Given a hypothesized close relationship between *Auroracetus* and *Pontoporia* in South America, the occurrence of *Auroracetus* on the eastern US coast represents dispersal to North America. Alternatively, if a closer affinity of *Auroracetus* to *Pliopontos* is discovered, then dispersal via the Central American (Panamanian) seaway before closure of the Panamanian isthmus is suggested.

From the North Sea, the late Miocene *Protophocoena minima* has been described, based on a fragmentary and worn skull, from the Netherlands and, like *Pontistes* sp., from the Gram Formation, Denmark. It is likely that pontoporiids reached the North Sea via the south, passing through the Parathethys Sea, which resulted from the collision of Africa with the Iberian Peninsula in southwestern Europe (see also chapter 4).

Delphinoidea (Stem "Dolphins"): Delphinidae (Dolphins) and Monodontoidae. Albireonidae was proposed in 1984 as a new family of delphinoids to include *Albireo whistleri*, described on the basis of one of the most complete fossil cetacean skeletons known from the late Miocene (8-6 million year ago) Almejas Formation, Cedros Island, Baja California. A second, geochronologically younger *Albireo* species, *A. savagei*, based on a partial skeleton, was described later from the late Pliocene (3-2 million years ago) Pismo Formation in central California. Albireonids were medium-sized delphinoids, approximately 2.5 m (9 ft) in length with a deep thorax, relatively short neck, and broad fore flippers. Although the albireonids have been linked to either kentriodontids or phocoenids, recent phylogenetic analysis positions them as basal delphinoids, a clade that also includes delphinids (dolphins) and monodontoids (narwhal, beluga, porpoises) (Geisler et al., 2011).

The Oceanic dolphins, or Delphinidae, are the most morphologically and taxonomically diverse of the cetacean families and include 17 genera and 37 extant species of dolphins, as well as killer whales (*Orcinus orca*) and pilot whales (*Globicephala* spp.). Most delphinids are small to medium in size, ranging from 1.5 to 4.5 m (4.9-14.7 ft). The teeth are homodont and conical (figure 3.29). The

giant among them, the killer whale, reaches 9.5 m (31 ft) in length. Although the Irrawaddy dolphin, *Orcaella brevirostris*, found only in the Indo-Pacific, has been regarded as a monodontid (with, belugas and narwhals), more recent morphological and molecular work suggests that this species is a delphinid. Molecular analyses find little resolution among subfamily groups and suggest polyphyly in the genus *Lagenorhynchus*. Molecular divergence times for delphinids have been estimated at 10-9 million years ago, although the deepest divergences based on the fossil record date to the late Miocene.

The oldest confirmed delphinid, *Eodelphinus kabatensis* (figures 3.30, 3.31), is from the latest Miocene (13-8.5 million years ago) of Japan. This calibration point is in general agreement with molecular divergence dates for delphinids. Paleobiogeographic reconstruction and divergence times suggest a likely origin and early diversification of delphinid odontocetes during the middle Miocene in the Pacific Ocean, with subsequent migration to the North Atlantic via the Central American seaway. Significant fossil delphinid specimens from the Pliocene of Italy, beginning in the late 1990s, documented by Bianucci and colleagues (e.g., Bianucci, 1996, 2013; Bi-

Figure 3.29. Delphinid (*Stenella attenuata*) skull illustrating tooth anatomy.

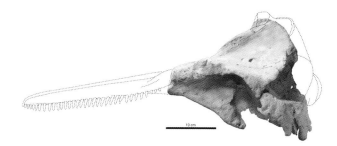

Figure 3.30. Skull of *Eodelphinus kabatensis*. Courtesy of M. Murakami.

Figure 3.31. Life restoration of *Eodelphinus kabatensis*. Illustrated by R. Boessenecker.

anucci and Landini 2002) corroborate the explosive radiation of delphinids during the Pliocene-Pleistocene. Eight genera have been described from this time interval. This diversification has been related to morphological disparity in delphinid species, including enhanced echolocation abilities and a related increase in brain size. Various hypotheses for this diversification have been proposed, including rapid recolonization of the Mediterranean after the Messinian salinity crisis. As a result, the flow of warm water from the Mediterranean to the Atlantic was shut off, which intensified the cooling of the North Atlantic. Resulting steep thermal gradients were probably associated with increased levels of upwelling and phytoplankton productivity in the North Atlantic.

One of the better-represented Pliocene delphinids from Italy is *Septidelphis morii*, previously referred to the extant Atlantic spotted dolphin *Stenella* cf. *S. frontalis*. This new fossil delphinid is based on a partial skeleton of an immature animal. The skull is relatively large compared with extant delphinids. Cladistic analysis positions this taxon as sister to another Pliocene Italian delphinid, *Etruridelphis*, with more distant affinities to *Stenella* (figure 3.32). Bianucci (2013) further suggests that extant genera

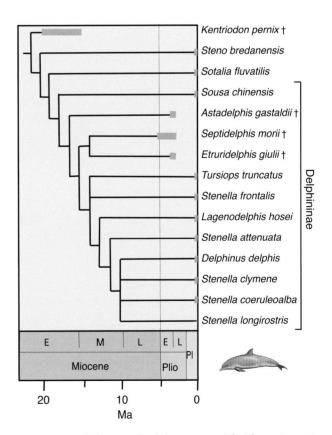

Figure 3.32. Phylogeny of Delphininae. Modified from Bianucci, 2013.

of delphinids probably did not arise until the Pleistocene and that most Pliocene delphinids are best allocated to extinct taxa. The taxonomy of extant delphinids is the source of continuing controversy, given the lack of consistency or resolution in molecular phylogenies.

Several fossil species have been related to the delphinid subfamily Globicephalinae. Globicephalines are also known as "blackfish" and include several modern species, such as the pilot whale, *Globicephala* spp., and false killer whale, *Pseudorca crassidens*. Extant species are characterized by dark coloration, large body size, blunt rostra, large melons, and low tooth counts. Estimates of the divergence of the Globicephalinae range from 8 to 4 million years ago, although the widespread diversity of globicephalines during the late Pliocene suggests that the 4-million-year estimate is too young. Fossil globicephalines of Pliocene age are known from the United States, Chile, England, Japan, Mexico, Netherlands, and Spain (Boessenecker, 2013). A new genus and species of dolphin, *Platalearostrum hoekmani*, was described from the Plio-Pleistocene. The partial skull was collected from dredging operations in the North Sea (Netherlands). A conspicuous feature of this species that it shares with extant *Globicephala* is a very short, broad rostrum with a rugose dorsal area for muscle attachments, which may have housed a large melon that extended beyond the rostrum (figure 3.33). A related species is *Protoglobicephala mexicana*, described on the basis of a skull from the late Pliocene of Baja California. *Hemisyntrachelus oligodon*, known from the North Sea and the Pisco Formation of Peru, was recently transferred from *Tursiops* to *Hemisyntrachelus*. This genus was reported to have globicephaline affinities.

Monodontoidae: Monodontidae (Narwhals), Phocoenidae (Porpoises), and ?Odobenocetopsidae. Traditionally, phocoenids and delphinids have been recognized as more closely related to one another than either is to monodontids, although a recent combined analysis of cetaceans (Geisler et al., 2011) supported monophyly of the Delphinoidea and an alliance of Monodontidae + Phocoenidae and possibly Odobenocetopsidae in a new clade, Monodontoidae.

Porpoises include seven small to medium-sized extant species living in subtropical to cool temperate shallow, coastal waters. Phocoenids are distinguished from other odontocetes by their flattened, spatulate-shaped rather than conical teeth (figure 3.34). The most recent analysis of phylogenetic relationships among extant taxa based on both morphological and molecular data positions the finless porpoise, *Neophocaena phocaenoides*, as the most basal extant member of the family. The most endangered cetacean, the vaquita, *Phocoena sinus*, is allied with Burmeister's porpoise, *Phocoena spinipinnis*, with *Neophocoena* sister to that clade. Dall's porpoise, *Phocoenoides dalli*, is sister taxon to *Phocoena dioptrica* and the harbor porpoise, *Phocoena phocoena*. Molecular data differ in supporting an unresolved relationship between *P. spinipinnis* and *P. sinus* and association of these two species with the spectacled porpoise, *Phocoena dioptrica*.

Like delphinids, phocoenids have a fossil record that extends back to the late Miocene and Pliocene in the North Pacific and North and South Atlantic. Thirteen genera and 15 fossil species have been described. The majority of fossil phocoenids are known from the Pacific, which supports a Pacific origin for the group, as does the occurrence of related delphinoids (e.g., Albireonidae, Delphinidae, Monodontidae, Odobenocetopsidae) along Pacific coasts. Extinct stem porpoises from Japan include *Pterophocoena* from the Miocene (9.3-9.2 million years ago), *Archaeophocoena* and *Miophocoena* from the late Miocene (6.4-5.5 million years ago), and the slightly younger *Numataphocoena* and *Haborophocoena* from the late early to early late Pliocene (Murakami et al., 2014). In contrast to molecular results that position phocoenids and monodontids as sister taxa, morphological data, especially the cranial anatomy of *Archaeophocoena* and *Miophocoena*, support a sister group relationship for delphinids and phocoenids. Later-diverging phocoenids include *Salumiphocaena*, from the late Miocene, California; *Piscolithax*, from the latest Miocene to early Pliocene, Mexico; and *Australithax* and *Lomacetus* from the late Miocene, Peru. Two phocoenids are known from outside the Pacific Ocean, *Septemtriocetus* and *Brabocetus* from the late Pliocene of the North Sea, which together with the Pleistocene-Recent occupation of the North Sea and North Atlantic

Figure 3.33. Life restoration of *Platalearostrum hoekmani*. From Post and Kompanje, 2010.

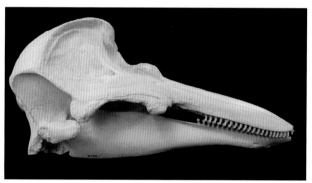

Figure 3.34. Phocoenid (*Phocoena phocoena*) skull illustrating tooth anatomy.

supports several trans-Arctic dispersal events to account for this distribution. A dispersal route across the Bering Straits during the early stages of its opening (5.4–4.8 million years ago) has been proposed for phocoenids, the walrus, phocid seals, and other marine organisms (molluscs, echinoderms, barnacles, eelgrass, kelp, red algae).

A fossil porpoise from the Pliocene (5.0–1.6 million years ago) of California, based on a skull and jaws and postcranial elements, was named *Semirostrum cerutti* for the shortened rostrum relative to the mandible (Racicot et al., 2014). This extinct porpoise is hypothesized to have used its long, fused, nearly edentulous mandibular symphysis—which extends farther beyond the ros-

trum than in any other known mammal (figure 3.35)—to probe for and obtain prey, probably small fish and cephalopods. Evidence for the tactile sensitivity of the lower jaw comes from CT images that show extensive elongate arterial canals that housed blood vessels supplying nourishment to soft tissues. Well-developed neck muscles in *Semirostrum*, as evidenced by deep dorsal condyloid fossae and robust, unfused neck vertebrae, permitted head mobility and maneuverability. Both extant and fossil porpoises exhibit developmental processes that suggest slow rather than rapid growth (see the following discussion), the mechanisms of which will require further testing.

The relatively short rostrum, small body, reduced fecundity, and longevity of extant phocoenids compared with delphinoids have been used to argue for paedomorphism, a developmental process characterized by retention of juvenile characteristics by adults. It has been suggested that major climatic fluctuations during the Pleistocene may have promoted the paedomorphic lifestyle of phocoenids as an adaptive strategy. The shortened ontogenies that result allow sexual maturity to be reached earlier and provide a greater reproductive potential (Galatius et al., 2011). The morphometrics of the skeleton of extant phocoenids examined by Galatius and

Figure 3.35. Schematic reconstruction of fossil porpoise *Semirostrum cerutti* based on skull and postcrania (*top*) and life restoration (*bottom*). Courtesy of R. Bossenecker.

colleagues has been further related to habitat and feeding, distinguishing between coastal and pelagic species.

Another study explored phocoenid inner ear anatomy in extant species in relation to ecology and body size, with results suggesting that species with offshore habitats (*P. dioptrica, P. dalli*) have semicircular canal measurements that indicate a vestibular system that is more sensitive to head rotation than that of coastal species (*P. phocoena, P. sinus, P. spinipinnis, Neophocaena phocaenoides*). Pelagic species are faster swimmers with quicker prey capture, whereas coastal species have a reduced sensitivity to head rotations that may be a consequence of their slower movements in scanning the environment and searching for prey. Extrapolations to fossil phocoenids suggest that species with inner ear measurements in the range of more-pelagic extant phocoenids include *Pterophocoena,*

Piscolithax, and *Salumiphocoena* (Racicot et al., 2016). Several other fossil phocoenids, notably *Numataphocoena, Haborophocoena,* and *Semirostrum,* have inner ear measurements that fall in the range of coastal extant phocoenids. These morphology results are confirmed by fossil locality data, since *Semirostrum* has been recovered from the San Diego Formation, a coastal depositional environment, and *Salumiphocoena* from the Monterey Formation, a pelagic to marginal marine environment. Given the excellent preservation of ear bones in the fossil record, the ability to infer ecology is especially informative.

Monodontids include two intermediate-sized extant species of toothed whales that at present occur only in the northern hemisphere: the narwhal, *Monodon monoceros,* and the beluga or white whale, *Delphinapterus leucas.* Except for the tusk, narwhals are edentulous. The presence of a spiraled canine tusk reaching 2.6 m (8.5 ft) in length distinguishes males, but is occasionally present in females. The narwhal tusk may be the source of the legend of the unicorn, a horse with cloven hooves, a lion's tail, and a horn in the middle of its forehead that closely resembles the narwhal tusk. The function of the tusk is still debated, but it is almost certainly a secondary sexual character employed in aggressive encounters or intraspecific display. New anatomical and physiological evidence suggests that it might also have a sensory function, perhaps detecting waters in which breeding females may be gathered.

The beluga is characterized by its completely white coloration, which affords protection from predators in the Arctic. Both the beluga and narwhal are endemic taxa and occupy Arctic waters. The beluga has a circumpolar distribution, whereas the narwhal occupies the North Atlantic and occurs as vagrant in the Pacific. *Delphinapterus* is known from abundant Pleistocene localities in upper New York State and western Canada. Other fossil monodontids include *Denebola* from the late Miocene Almejas Formation of Baja California and *Bohaskaia* (figure 3.36) from the North Atlantic early Pliocene. During the late Miocene and Pliocene, monodontids occupied temperate waters as far south as Baja California, which indicates that extant monodontids evolved under different environmental conditions and that their Arctic cold-adapted distribution is a more recent event.

Odobenocetopsidae. An extinct relative of the monodontids, the bizarre cetacean *Odobenocetops* is placed in its own family, Odobenotopsidae. In its morphology and inferred benthic suction-feeding habits it is convergent with the modern walrus (figure 3.37). As in the walrus, the snout and inferred vibrissae and the deep palate suggest that *Odobenocetops* fed on benthic invertebrates such as molluscs or crustaceans. *Odobenocetops* is known by two species from the early Pliocene (4–3 million years ago) Pisco Formation of Peru, both of which exhibited sexual dimorphism. The geographic restriction of *Odobenocetops* to Peru is an excellent example of endemism. Among extant cetaceans, narwhals, beluga, and bow-

Figure 3.36. Life restoration of *Bohaskaia monodontoides* with extant narwhal and beluga in background. Illustrated by C. Buell.

heads are endemic species that are now limited to Arctic waters, although they occurred further south in warmer waters during the late Pliocene. In *Odobenocetops*, the rostrum is short, rounded, and blunt, formed by greatly enlarged premaxillae, differing from the elongate rostrum of other cetaceans. The eyes were relatively large and oriented upward, indicating binocular vision, rather than to the sides of the body as in most other dolphins. A distinctive feature of *Odobenocetops* is the posteroventrally oriented tusks. In the only known male specimen of *O. leptodon*, the erupted portion of the needle-like right tusk was 1.07 m (3 ft 6 in) long, whereas the small left tusk was only 25 cm (9.8 in) (Muizon et al., 2001). Given their thinness, the tusks probably played no role in feeding but more likely served for social display among males. Body length of *Odobenocetops* has been estimated at 3–4 m (9.8–13.1 ft). Large, well-developed occipital condyles suggest that it had strong, flexible neck musculature and swam with its head bent ventrally and parallel to the body.

Figure 3.37. Life restoration of *Odobenocetops leptodon.* Illustrated by M. Parrish.

CROWN ODONTOCETES OF UNCERTAIN RELATIONSHIPS

The phylogenetic position of Eurhinodelphinids, long-beaked extinct "dolphins" (previously known as Rhabdosteidae), is uncertain at present. Eurhinodelphinids have been considered stem group to Ziphiidae or sister group to various other stem clades, such as Squalodontidae + Squalodelphinidae (Geisler et al., 2011) or Platanistoidea + Delphinida + Ziphiidae (Lambert et al., 2015b). They were geographically widespread (ranging from the North Sea to Australia) and moderately diverse (at least eight genera known) during the early and middle Miocene (23-7 million years ago) and became extinct in the late Miocene.

Evolutionary trends seen in the later-diverging *Xiphiacetus* compared with more basal taxa (e.g., *Eurhinodelphis*) include structures related to telescoping of the skull (progressive movement of nares posteriorly), hearing (better isolation of the ear bones), and feeding apparatus (reduction in size of the temporal fossa related to size and type of prey). Eurhinodelphinids are hypothesized to have been slow-swimming, shallow-water dwellers. One late Oligocene species was found in freshwater deposits in Australia. The edentulous anterior part of the rostrum, longer than the mandible, is suggested to have functioned as an efficient probing/foraging tool. Eurhinodelphinid relationships are contentious. Most recently, they have been included either in a clade with ziphiids, delphinids, or platanistids or in a clade between physeteroids and other crown odontocetes.

In addition to Eurhinodelphinidae, other families of long-snouted dolphins of uncertain affinities originating or diversifying during the early Miocene include the Allodelphinidae, "Dalpiazinidae," and Eoplatanistidae.

As reported by Lambert and colleagues (2015b), a new long-snouted dolphin, *Chilcacetus cavirhinus*, based on several well-preserved skulls and associated jaws from the lower Miocene Chilcatay Formation of Peru and previously classified in the Eurhinodelphinidae, cannot as yet be assigned to any named odontocete clade. This species differs from eurhinodelphinids in possessing an unfused mandibular symphysis. Cladistic analysis of *Chilcacetus* supports a clade including *Chilcacetus, Macrodelphinus, Argyrocetus* from Argentina, and *"Argyrocetus"* from California, although statistical support and unambiguous synapomorphies are lacking.

Another problematic taxon is "Dalpiazinidae," proposed by Muizon (1988). It includes a single genus and species, *Dalpiazina ombonii*, known from the early Miocene of Italy and described as having a long rostrum with homodont dentition. This taxon also has been related to the Platanistoidea. The Eoplatanistidae, named for the genus *Eoplatanista* from the early Miocene, has most recently been allied with Eurhinodelphinids.

Baleen Whales: Mysticetes

Baleen whales include the largest animals on earth. Much of their success is due to their gigantic mouths and their ability to gulp and filter an enormous volume of water and prey during bulk feeding. There are 14 extant species of mysticetes, and more than 40 extinct species have been described in just the past 15 years.

Baleen and the Origin of Bulk Feeding

The name Mysticeti means "mustached whales," so named for their feeding apparatus. Baleen is a unique mammalian structure consisting of tough epidermal keratinous tissue organized into plates that are suspended from the roof of the mouth (figure 3.38). Throughout the life of the whale, its baleen plates grow continuously at their base and are worn down by the tongue, exposing a matted network of bristles. The frayed bristles on the medial margin of each plate overlap those of adjacent plates to produce a sieve that entraps prey within the mouth. The number and length of individual baleen plates on either side of the mouth vary from approximately 155 plates, the longest

Figure 3.38. Baleen positioned in the mouth of a humpback whale during lunge feeding. Courtesy of A. Friedlander.

measuring 0.4–0.5 m (1–1.5 ft), in gray whales (Eschrichtiidae) to more than 355 plates, some exceeding 3 m (9 ft) in length, in right whales (Balaenidae). Mysticetes use their baleen racks in combination with other unique anatomical and behavioral specializations to capture aggregations of small fish or invertebrates or both. Extant adult mysticetes possess baleen and lack teeth as adults; they have deciduous teeth in the fetal stage that degenerate and are resorbed before birth.

The transition from teeth to baleen and raptorial feeding to bulk feeding can be traced in the fossil record. The origin of bulk feeding represents a major morphological and ecological shift in mammalian evolution. This novel filter-feeding strategy is a key innovation that heralded the evolution of modern baleen whales. But bulk feeding did not begin with mysticetes. The bulk-feeding niche exploited today by baleen whales was probably occupied in the past by now extinct giant bony fish during the Mesozoic (270–86 million years ago) and plankton-feeding sharks and rays in the Cenozoic (66–23 million years ago), using gill rakers instead of baleen as filters. Other similarities shared by these groups include changes in dentition and mandibular geometry (elongate jaws reflecting a reduction in force transmission and a less powerful bite), loss of teeth, and evolution of giant body size. The demise of many of these large-bodied marine fish at the end of the Cretaceous most likely created an ecological opportunity for the evolution of suspension-feeding sharks and rays in the middle Eocene and filter-feeding mysticetes in the late Eocene and Oligocene (figure 3.39).

Fossil baleen is rare in the fossil record because baleen does not have a large mineral content and deterio-

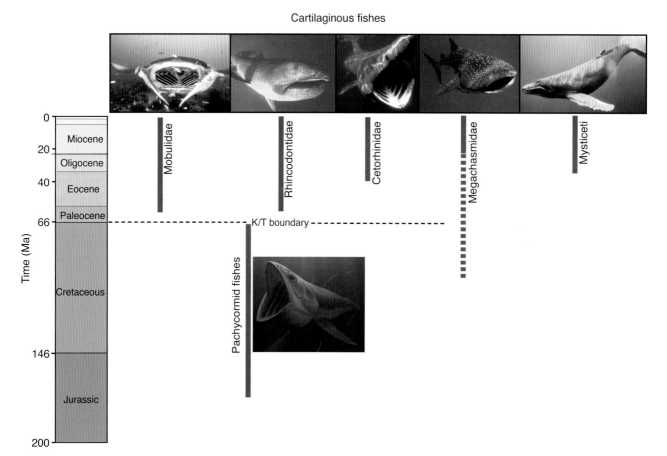

Figure 3.39. Ecological replacement of various fish lineages with baleen whale. *Abbreviation:* K/T, Cretaceous/Tertiary boundary. Modified from Friedman et al., 2010.

rates soon after death. Occasionally, however, the presence of hydroxyapatite (the mineral component of bone) and traces of other elements (e.g., manganese, copper, iron, and calcium) in the baleen of fossil whale skeletons allowed preservation in sediments with rapid mineralization and burial, such as in the Miocene-Pliocene Pisco Formation of Peru. The preserved baleen appears as a series of colored plates that contrast with the surrounding diatomaceous mudstone. Observations of modern whale carcasses on the seafloor and of stranded individuals indicate that the baleen typically detaches from the mouth of the whale after death. In contrast, the exceptionally well-preserved baleen in the Pisco Formation mysticetes is often found in life position, suspended from the mouth. This exceptional preservation of baleen indicates the rapid burial of the whale carcasses in the Pisco Basin (figure 3.40).

Although baleen rarely fossilizes, bony vascular structures on the palate of toothless mysticetes (30 million years ago to present day) are hypothesized as osteological correlates for the presence of baleen. Archaic baleen whales include both toothed and toothless forms and are

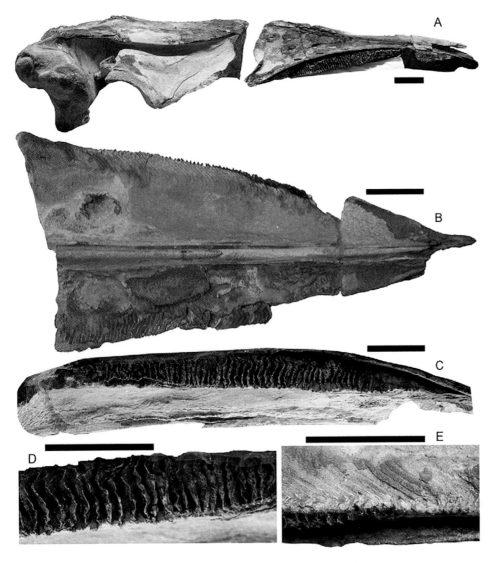

Figure 3.40. Skull of *Miocaperea pulchra* showing fossil baleen. (*A*) skull, lateral view, (*B*) rostrum, ventral view, (*C*) rostrum in right ventrolateral view, showing entire baleen series as preserved, (*D*) close-up of baleen series in lateral view, (*E*) close-up of left baleen series in ventral view (ventral is upside). Scale bars = 50 mm (1.96 in). From Bisconti, 2012.

separated in the fossil record by a gap in anatomy. An evolutionary solution to closing this gap in feeding anatomy is revealed by the recent discovery of tiny openings on the palate between the teeth in 34- to 24-million-year-old toothed mysticetes (e.g., *Aetiocetus, Fucaia, Morawanocetus*). These tiny lateral palatal foramina and associated sulci are similar to those present in extant mysticetes and are hypothesized to function for the passage of blood vessels that nourish baleen (figure 3.41). A recent report reveals that baleen receives the majority of its blood from the superior alveolar artery and its associated branches (major distributors of blood to the teeth of toothed mammals) that pass through lateral nutrient foramina (Ekdale et al., 2015). The simultaneous occurrence of teeth and vascular structures implies that both teeth and baleen were present in stem mysticetes (figures 3.41, 3.42).

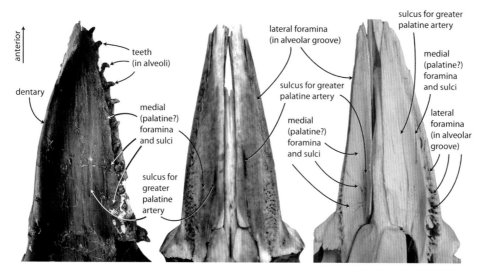

Figure 3.41. Palates of select mysticetes. *Left to right* (all in ventral view): extinct toothed mysticete (*Aetiocetus weltoni*), juvenile gray whale (*Eschrichtius robustus*), fetal fin whale (*Balaenoptera physalus*). From Ekdale et al., 2015.

Figure 3.42. Reconstruction of *Aetiocetus* with baleen and teeth. From Deméré et al., 2008. Illustrated by C. Buell.

However, as noted by some scientists (Marx et al., 2015, 2016b), even if aetiocetids had some form of baleen precursor, it does not necessarily follow that the latter was employed in filter feeding. It has been proposed instead that some aetiocetids may have had heightened gums, as suggested for mammalodontids based on their emergent tooth crowns, implying well-developed gingival tissue. Although enlarged gum tissue might have better allowed a biting, mostly raptorial suction-feeding strategy, this hypothesis remains untested.

The origin of baleen appears to have been a stepwise transition from an ancestor with teeth only, to an intermediate stage with functional teeth and baleen, to the derived condition with baleen only. Numerous anatomical specializations allow extant mysticetes to process a large volume of water and facilitate bulk filter feeding with baleen. Among such features are a broadened rostrum, thin maxillae, lateral bowing of the mandibles to accommodate the baleen plates, and an unsutured mandibular symphysis that allows the lower jaws to move independently.

TOOTH-TO-BALEEN TRANSITION: GENETIC EVIDENCE

Molecular studies of tooth genes important in the development of tooth enamel show that mutations of these genes are associated with dental defects in which the enamel is thin and malformed. Since toothless mysticetes descended from ancestors with fully mineralized teeth, scientists hypothesized that enamel-specific genes would be present but nonfunctional in the mysticete genome. This prediction was confirmed by the discovery of frameshift mutations in the dental genes. In a frameshift mutation, one or more bases are inserted or, as in this case, deleted, which disrupts the reading frame of the dental genes and suggests that these loci are decaying enamel-specific pseudogenes (figure 3.43). Thus, the end result in this evolutionary transition from teeth to baleen

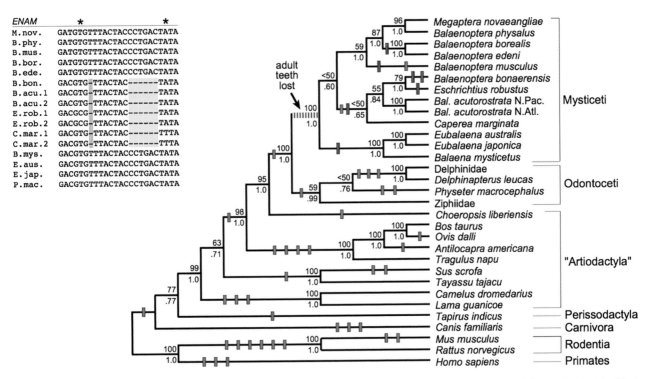

Figure 3.43. Frameshift mutation of *AMBN* dental gene, showing aligned sequences for five mysticetes and domestic pig. Modified from Deméré et al., 2008.

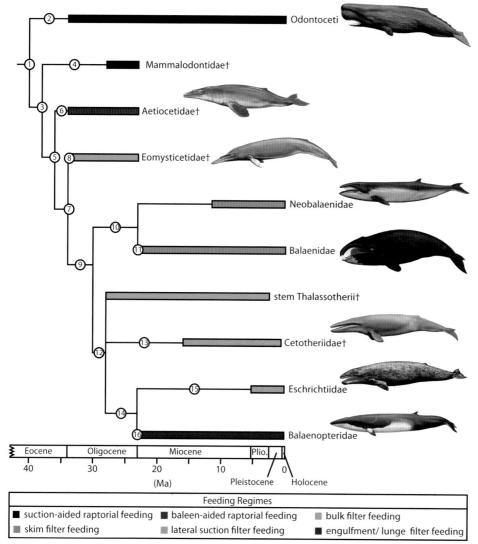

Figure 3.44. Cladogram with associated feeding strategies in extinct and extant mysticetes. Relationships are based on a strict consensus of cladograms presented by Boessenecker and Fordyce (2015c). Stratigraphic ranges were downloaded from the Paleobiology Database (paleobiodb.org) on June 9, 2016, by searching for geologic strata through the taxonomic name search form for members of each terminal taxon named on the cladogram. (1) CETACEA: 28%–59% olfactory genes nonfunctional. (2) ODONTOCETI: 74%–100% olfactory receptor genes nonfunctional; biosonar ("echolocation"). (3) MYSTICETI: broad rostrum; thin lateral margins of maxilla; small, wide teeth with denticles; diastemata in upper jaw. (4) MAMMALODONTIDAE: enlarged, forward-facing eyes (enhanced visual acuity?). (5) AETIOCETIDAE + CHAEOMYSTICETI: small lateral nutrient foramina (presence of baleen-like structure); mandibular symphysis unfused. (6) AETIOCETIDAE: enlarged, forward-facing eyes (enhanced visual acuity?). (7) CHAEOMYSTICETI: low ^{13}C in bones (feeding at low trophic level); reduction of postnatal dentition; long, flattened rostrum; premaxilla/maxilla kinesis; elaborate lateral nutrient foramina; reduced function of tooth and enamel genes. (8) EOMYSTICETIDAE: vestigial anterior dentition. (9) CROWN MYSTICETI: definitive presence of baleen/postnatal teeth absent; anterior extension of occipital shield; lateral bowing of dentaries; reduced function of tooth and enamel genes. (10) BALAENOIDEA: arching of rostrum; long baleen; subrostral gap. (11) BALAENIDAE: extreme arching of rostrum. (12) THALASSOTHERII: anteriorly convergent baleen racks; indistinct/concave glenoid fossa; loss of *C4orf* tooth genes (?). (13) CETOTHERIIDAE: reduced gaping of mouth; semi-synovial jaw joint. (14) BALAENOPTEROIDEA: loss of *C4orf* tooth genes. (15) ESCHRICHTIIDAE: slight arching of rostrum; short, coarse baleen; semi-synovial jaw joint; few, short throat grooves. (16) BALAENOPTERIDAE: synovial jaw joint lost; tongue reduced and flaccid; ventral throat pouch; mandibular sense organ; elongation and multiplication of throat grooves. From Berta et al., 2016. Illustrations of whales by C. Buell.

is the presence of vestigial genes in the mysticete genome that represent "molecular" fossils and provide genetic evidence of the toothed ancestry of baleen whales.

The fossil record of mysticetes provides anatomical and molecular evidence for the tooth-to-baleen transition. A summary of morphological and molecular transformations mapped onto a cladogram is presented in figure 3.44 (Berta et al., 2016).

FEEDING HABITS: ISOTOPE EVIDENCE

Dietary information can be obtained from the isotopic concentration of carbonates in the bones and teeth of marine mammal consumers (see also chapter 1). This information can be used to reconstruct the migratory movements and foraging habits of extant mysticetes and to explore these behaviors in the fossil record. Toothed mysticetes (Mammalodontidae) and odontocetes have similar ^{13}C values in bones and teeth, which suggest feeding on fish and squid and do not suggest filter feeding or feeding low in the food chain. In contrast, ^{13}C values for eomysticetids and stem balaenopteroids are lower and significantly different from those of toothed mysticetes or odontocetes and suggest that these mysticetes may have had a specialized diet of zooplankton, similar to extant mysticetes (figure 3.45). Values of ^{13}C for kekenodontids ("archaeocetes" closely related to Neoceti; see chapter 2) were also found to be low and showed the greatest range, overlapping with those of both edentulous mysticetes and resident species (i.e., those living at a particular latitude). Clementz and colleagues (2014) hypothesized that these results correlate with tooth type, perhaps reflecting an ontogenetic shift in diet associated with the transition from nursing as neonates to eating marine prey as adults. The evolution of bulk feeding would have allowed large consumers to take advantage of short but regular bursts of an abundant food supply.

Skull Growth and Development

Ontogenetic changes in the vertebrate skull have numerous functional, ecological, and behavioral consequences. Relatively little is known, however, about the growth and development of the mysticete skull. Tsai and Fordyce

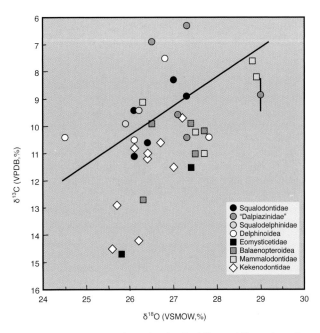

Figure 3.45. Bivariate plot of individual ^{13}C and [18]O values for Oligocene fossil cetaceans from the United States and New Zealand. *Solid line* is regression for modern cetacean isotope values after correction for changes in atmospheric CO_2. *Abbreviations:* VPDB, accepted standard for carbon; VSMOW, accepted standard for oxygen. From Clementz et al., 2014.

(2014) used geometric morphometrics (2-D and 3-D analysis of shape variation) to investigate the ontogeny of the skull of several mysticetes, ranging in age from fetus to subadult/adult—two balaenopterids, the sei whale, *Balaenoptera borealis*, and the humpback, *Megaptera novaeangliae*, and the pygmy right whale, *Caperea marginata* (figure 3.46). The analysis indicated that the two clades of mysticetes show different developmental processes. Relatively few changes occur during the ontogeny of *Caperea*, with juvenile and adult specimens plotting in the same region of morphospace. However, adult *Balaenoptera* and *Megaptera* plotted closely, with juveniles of each plotting close together but not with the adults. This indicates not only more extreme ontogenetic changes (generally reflecting the extreme lengthening of the rostrum) in balaenopterids relative to *Caperea*, but also similar ontogenetic trajectories in each rorqual. Tsai and Fordyce thus show *Caperea* as undergoing paedomorphosis whereas the two rorquals undergo substantially more changes during

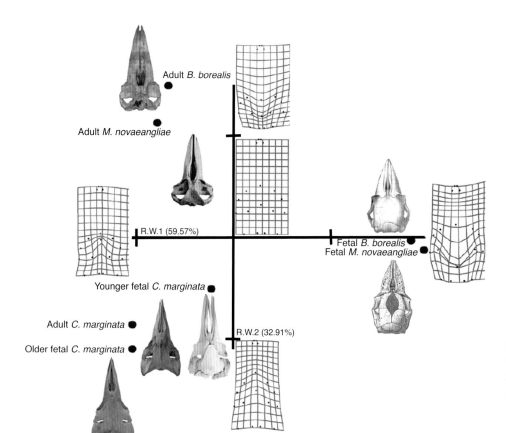

Figure 3.46. Shape variation in the skull of mysticetes, analyzed by geometric morphometrics. *Abbreviations:* R.W.1, relative warp axis 1; R.W.2, relative warp axis 2. From Tsai and Fordyce, 2014.

ontogeny than the hypothetical ancestral mysticete, thus undergoing peramorphosis (accelerated growth). The authors further suggest that the release of developmental constraints may have led to low taxonomic diversity in the single surviving neobalaenid and higher diversity among balaenopterids. Further study of skeletal growth is necessary in these and other mysticetes, especially balaenids and eschrichtiids and with larger sample sizes, to elucidate the relationship between developmental patterns and taxonomic diversity.

Stem Toothed Mysticetes: Llanocetidae, Mammalodontidae, and Aetiocetidae

Stem toothed mysticetes have been grouped into three families: Llanocetidae, Mammalodontidae, and Aetiocetidae (figure 3.47). Mysticetes probably originated in the southern hemisphere, since most stem mysticetes occur there.

The geologically oldest described mysticete is *Llanoce-*

tus denticrenatus, the only member of its family, Llanocetidae. Although it is known from a fragment of a lower jaw and associated skull, only the lower jaw and teeth have been formally described (the skull is still under study). Both fossils were collected from the late Eocene or early Oligocene (ca. 34 million years ago) marine sediments on Seymour Island, Antarctica. *Llanocetus* was a large, toothed whale with a skull length of about 2 m (6.5 ft). In addition to widely spaced multicusped teeth, *Llanocetus* had fine grooves around the tooth alveoli that indicate the presence of a blood supply to the palate, which is suggestive that it also had baleen.

Three other archaic toothed mysticetes, *Mammalodon colliveri*, *Mammalodon hakataramea*, and *Janjucetus hunderi*, placed in the Mammalodontidae (figure 3.44, node 4), were described by Australian vertebrate paleontologist Erich Fitzgerald (2006, 2010, 2012a) and Fordyce and Marx (2016). These mysticetes are known from late Oligocene and early Miocene, 28- to 24-million-year-old rocks

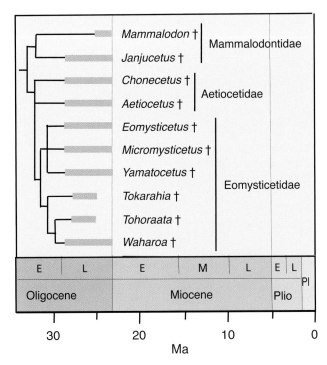

Figure 3.47. Phylogeny of stem mysticetes. From Boessenecker and Fordyce, 2015a.

of Australia and New Zealand. *J. hunderi* is distinguished by its short, triangular snout, heterodont teeth, and large orbits. The large eyes, proportionately larger than those of other toothed mysticetes, indicate that vision was well developed in this stem mysticete (figures 3.44, 3.48). Based on the size of the skull, *Janjucetus* was small, with a body length of 2.9-3.2 m (9.5-10.5 ft). The robust jaws, large teeth, and tooth wear facets suggest that *Janjucetus* was a raptorial predator that ate single prey items such as fish. With its relatively broad rostrum, indicating initial enlargement of the oral cavity, but its sutured jaws, *Janjucetus* was viewed by Fitzgerald (2012a) as occupying a transitional step in the evolution of bulk feeding in later-evolving edentulous mysticetes, with mandibular rotation on their long (α) and horizontal (Ω) axes and kinetic jaws.

Mammalodon has a short, broad rostrum with reduced slender incisors and multicusped cheek teeth (figure 3.49). Although originally hypothesized as a lunge filter-feeder, *Mammalodon*'s dentition, rostrum, skull, mandible, and sternum suggest that it more likely used suction during prey capture. For example, its blunt jaws are

similar to those of odontocete species known to employ oral suction in feeding. The manubrium and sternum are specialized with an enlarged surface area for the origin of a hypertrophied sternohyoideus muscle, known to be a key component of suction feeding. Its relatively large eyes were dorsoventrally directed. *Mammalodon* is among the smallest toothed mysticetes, with a skull length of 45 cm (1 ft 5.1 in). The small body size, relatively large orbits, low cranium, and relatively large occipital condyles are interpreted as having evolved through paedomorphism, as seen in other mysticetes (e.g., *Caperea*).

Contemporaneous with mammalodontids of the southern hemisphere are the Aetiocetidae, a monophyletic lineage of toothed mysticetes well documented in the northern hemisphere. These are often described as bridging the morphological gap between stem toothed mysticetes and baleen-bearing mysticetes. Aetiocetids, which include five genera and eight species (*Aetiocetus, Chonecetus, Ashorocetus, Fucaia, Morawanocetus*), are known from Oligocene rocks on both sides of the North Pacific. Another taxon, *Willungacetus aldingensis* from South Australia, provisionally referred to this family, requires further study before it can be confirmed as an aetiocetid. Additional undescribed aetiocetids have been reported from California and Baja California. The oldest reported aetiocetid, *Fucaia buelli*, was described by Felix Marx and colleagues (2015) from the earliest Oligocene (ca. 35-31 million years ago) of western North America (Washington State). Of particular significance is that this taxon fills a temporal gap between aetiocetids and the oldest known mysticete, *Llanocetus*.

The most speciose genus is *Aetiocetus*, with four species: *A. tomitai* and *A. polydentatus* from Japan and *A. weltoni* and *A. cotylalveus* from Oregon. Recent phylogenetic analyses position *A. cotylalveus, A. polydentatus,* and *A. weltoni* more crownward relative to other aetiocetid taxa. The best known aetiocetid is *A. weltoni*, about 3-3.4 m (10-11 ft) in length based on skull length. Vision was well developed as judged from the large orbits (figure 3.44, node 6). As previously discussed, the most notable feature hypothesized for *A. weltoni* is palatal vascular structures suggestive of its possessing both baleen

Figure 3.48. Life restoration of *Janjucetus hunderi*. Illustrated by C. Buell.

Figure 3.49. Life restoration of *Mammalodon colliveri*. From Fitzgerald, 2010. Illustration by B. Choo.

and a functional adult dentition. Several features of the skull and mandible (e.g., long rostrum and lower jaws, large coronoid process) suggest that *A. weltoni* was capable of a snapping style of jaw adduction similar to that of dorudontine "archaeocetes," rather than sustained mandibular abduction as seen in extant mysticetes. Although originally placed in the Archaeoceti, *Chonecetus*, which currently consists of a single species, *C. sookensis* from British Columbia, is now recognized as an aetiocetid. In addition to *Morawanocetus yabukii* described from the Oligocene (26–24 million years ago) Morawan Formation of Japan, several as yet undescribed species dating from 19–17 million years ago are known from California. Also described from the Morawan Formation of Japan is *Asho-*

rocetus eguchii, based on the posterior part of a cranium. A new aetiocetid reported from Washington exhibits tooth wear that has been interpreted to indicate that suction rather than raptorial feeding was used for prey capture (Marx et al., 2016b). Further study of tooth wear in other aetiocetids is needed to confirm this interpretation. Aetiocetids had a rigid skull similar to that of mammalodontids, but unlike them, aetiocetids had an unfused mandibular symphysis indicating a kinetic lower jaw.

Nearly all aetiocetids are reported to be of small body size, ranging from 3 to 4 m (9.8–13 ft). *Fucaia buelli*, at an estimated body length of 2.1–2.2 m (~7 ft), is one of the smallest mysticetes. However, discovery of a *Morawanocetus*-like ear bone from the Morawan Formation of Japan indicates that some aetiocetids had a large body, as the bone was estimated to have come from an animal nearly four times as large (~8 m [26 ft]) as currently known aetiocetids (figure 3.50). Further, Tsai and Ando (2015) suggest that this large aetiocetid body indicates the niche partitioning of aetiocetids, implying different food resources and feeding strategies, rather than indicating sexual dimorphism. Although largely speculative, this hypothesized structure of aetiocetid communities is worthy of further consideration. In another study that reconstructed ancestral size among stem mysticetes onto phylogeny, Tsai and Kohno (2016) show that mysticetes originated from predecessors with a small rather than a large body as previously proposed. Their results further suggest that stem mysticetes evolved toward large body size multiple times independently. In extant mysticetes there is limited sexual dimorphism, with females often slightly (5%) larger than males, which has been related to their larger energetic demands, especially when carrying a fetus. By contrast, in extant odontocetes, males are generally larger than females, with the sperm whale exhibiting extreme dimorphism: males are more than two-thirds larger than females. Similarly, in the protocetid "archaeocete" *Maiacetus innus*, males were 12% larger than females.

In addition to these named toothed mysticetes, several other large toothed forms informally called "archaeomysticetes," from the Oligocene of South Carolina, are housed in the Charleston Museum and await formal description, though they have already been incorporated into phylogenies. Further study may position these taxa outside the Mysticeti, or even outside crown Cetacea.

Chaeomysticeti: Eomysticetidae

Baleen-bearing mysticetes, also known as the Chaeomysticeti, include several extinct lineages (figure 3.44, node 7). A newly described taxon from the late Oligocene of the North Pacific, *Sitsqwayk cornishorum*, is currently recognized as the most phylogenetically basal chaeomysticete (Peredo and Uhen, 2016). *Sitsqwayk* exhibits a mosaic of aetiocetid and eomysticetid features. The majority of early edentulous chaeomysticetes are members of the Eomysticetidae. The Eomysticetidae probably had a worldwide distribution, as they are now known in North America, Japan, and New Zealand. Divergence of the Chaeomysticeti has been estimated at 30.35 million years ago (Marx and Fordyce, 2015).

Eomysticetus whitmorei is known from the late Oligocene of South Carolina. The skull length is approximately 1.5 m (5 ft), which suggests that *E. whitmorei* was about the size of a minke whale, 7 m (23 ft) in body length. The skull indicates that *Eomysticetus* had a large temporalis muscle, which functions in jaw closing, but lacked the specialized elastic ligament seen in extant balaenopterids, which stores energy from opening the jaw that is then used to close it. The lower jaw of *Eomysticetus* is slender and straight, and this animal had not developed the laterally curved jaws of modern balaenopterids, which facilitate engulfment feeding. The presence of a large mandibular foramen in the lower jaw, a common feature in stem whales (see chapter 2), indicates that *Eomysticetus* was able to hear underwater sounds. Preserved flipper elements similar to those of extant whales are indicative

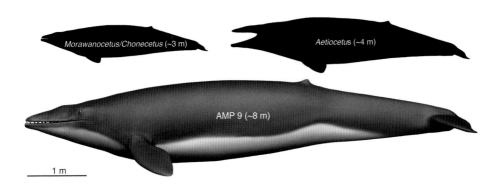

Morawanocetus/Chonecetus (~3 m)

Aetiocetus (~4 m)

AMP 9 (~8 m)

1 m

Figure 3.50. Life restoration of large *Morawanocetus*-like aetiocetid (Ashoro Museum of Paleontology) from Japan, compared with sizes of other aetiocetids. Tsai and Ando, 2015.

Figure 3.51. Life restoration of *Tohoraata raekohao* with skull superimposed. From Boessenecker and Fordyce, 2015a.

of limited mobility at the elbow joint and use of the flipper in steering.

An eomysticetid from the late Oligocene of Japan, *Yamatocetus canaliculatus*, consists of a nearly complete skull and lower jaws, several vertebrae and ribs, and forelimb elements. Both upper and lower jaws retain an alveolar groove, which suggests that it may have possessed teeth (although none were preserved) in addition to baleen.

A diverse southern hemisphere clade of eomysticetids including *Tohoraata, Tokarahia, Waharoa,* and *Matapanui* has recently been reported from the Oligocene of New Zealand, as documented in a series of papers by Boessenecker and Fordyce (2015a, b, c; 2016). One of the oldest well-dated fossils from the upper Kokoamu Greensand (29.8–27.3 million years ago) is named *Matapanui waihao,* from the Maori for "face," referring to the flat, planar frontal shield. *Matapanui* is currently recognized as the earliest-diverging member of the New Zealand eomysticetid clade. Another new genus is *Tohoraata,* which in Maori means "dawn whale" (figure 3.51), referring to the geologic age and archaic anatomy of this extinct whale. *Tohoraata* is currently known by two species: the slightly older *T. raekohao* (27–26 million years ago) collected from the Kokoamu Greensand and *T. waitakiensis* (26–25 million years ago) from the Otekaike Limestone. Other eomysticetids have a very small braincase and large brain cavities, attachment surfaces for jaw-closing muscles, and primitive ear bones. *T. raekohao* is a relatively small-bodied mysticete (~5 m in length). The species name *raekohao* comes from the Maori for "holes in the head," with reference to openings near the orbits that are also present in the "archaeocete" *Kekenodon onamata* (see chapter 2), of unknown homology and function.

A related new eomysticetid, *Tokarahia kauaeroa,* is described from the late Oligocene Otekaike Limestone of New Zealand, based on a well-preserved skull and partial skeleton. The problematic taxon *"Mauicetus" lophocephalus* is transferred to this new genus and recombined as *Tokarahia lophocephalus.* A putative tooth associated with the palate of *T.* cf. *T. lophocephalus* was recovered, suggesting the possible presence of adult teeth in this taxon. Referred material suggests that *T. kauaeroa* and *T. lophocephalus* co-occurred and were perhaps sympatric. The postcranial skeleton of *Tokarahia kauaeroa* is one of the most complete among the Oligocene mysticetes (figure 3.52), consisting of a mosaic of derived and ancestral features. For example, the elongate cervical vertebrae of *Tokarahia* resemble those of basilosaurids, whereas the immobile elbow joint is similar to that of extant mysticetes.

Another new eomysticetid, *Waharoa ruwhenua,* a genus name that means "long mouth" in Maori, is described from the Otekaike Limestone based on a well-preserved ontogenetic series of skulls and partial skeletons. This taxon provides evidence for the phylogenetically earliest

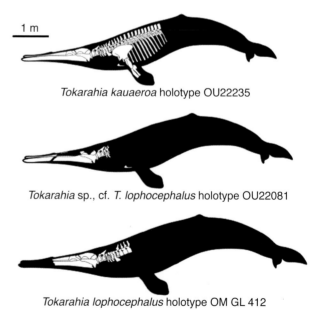

Tokarahia kauaeroa holotype OU22235

Tokarahia sp., cf. *T. lophocephalus* holotype OU22081

Tokarahia lophocephalus holotype OM GL 412

Figure 3.52. Silhouetted skeletal reconstructions of several *Tokarahia* skeletons. From Boessenecker and Fordyce, 2015b.

occurrence of a kinetic rostrum among mysticetes. Basilosaurids and most toothed mysticetes have closed or fused rostral sutures, indicating an immovable rostrum, in contrast to extant mysticetes in which the rostral elements are loosely articulated and open. Although associated teeth were not recovered, the presence of alveoli on the anterior palate and lateral foramina posteriorly suggests that both teeth and baleen may have been present (figure 3.53). Compared with the relatively short rostra of stem toothed mysticetes (aetiocetids, mammalodontids), rostral elongation characterizes the eomysticetids. And based on the more rapid rate at which the rostrum elongates in the ontogenetic series of *Waharoa*, the developmental process of accelerated growth (peramorphosis) may be responsible (Boessenecker and Fordyce 2015c). The preservation of juvenile specimens of *W. ruwhenua*, several under one year of age, suggests that the continental shelf of New Zealand was perhaps a calving ground during the Oligocene. Isotope samples from the tooth of another eomysticete, *Tokarahia*, are consistent with latitudinal migration, and it is possible that these eomysticetids, like other southern hemisphere marine mammals (i.e., seals), may have undertaken seasonal migrations to productive Antarctic waters, retreating to breed along the

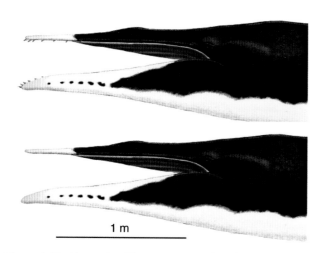

Figure 3.53. Alternative life restorations of *Waharoa ruwhenua* with (*top*) and without (*bottom*) dentition. From Boessenecker and Fordyce, 2015c.

New Zealand coast. Further isotopic study is needed to test this hypothesis.

Given that *Waharoa* had vascularization along the posterior three-fourths of its palate, its baleen apparatus may have been more limited in extent than in modern mysticetes. Other features such as the long rostrum, delicate lower jaw, and synovial jaw joint shared by *Waharoa* and other eomysticetids indicate that these marine mammals were not able to lunge-feed as do extant balaenopterids; rather, they may have employed a style of skim feeding similar to that of extant right whales. Like balaenids, they may have possessed a subrostral gap, although one formed by the absence of baleen rather than by laterally splayed racks as in balaenids. Some eomysticetids (*Yamatocetus*, *Waharoa*) possess alveoli or putative teeth (*Tokarahia*), but the size and shallow roots suggest they would not function well in filter feeding. It is also possible that teeth present in some eomysticetids may be vestigial or retained for social display (as in, e.g., ziphiids). As sister taxon to the crown group and with balaenids in the crown, the ancestral mode of feeding among these stem chaeomysticetes was probably a rudimentary form of skim feeding.

A new genus and species of stem edentulous mysticete, *Whakakai*, has been described from the late Oligocene of the Kokoamu Greensand (Chattian) of North Otago, New Zealand, by Tsai and Fordyce (2016). The larger, distinct periotic of *Whakakai* suggests that it may have provided a different style of receiving and/or processing sound. Specifically, the stylomastoid fossa and peribullary sinus, implicated in mysticete hearing, may have extended onto the squamosal. In the most recent phylogenetic analysis, *Whakakai* is allied with *Horopeta* in a clade that is sister group to crown Mysticeti. *Horopeta umarere*, which means "gulp or swallow" in Maori, has been described as a new genus and species of baleen whale from the early Oligocene (27-25 million years ago) of New Zealand (Tsai and Fordyce, 2015). *H. umarere* has a laterally bowed lower jaw and other features (e.g., a laterally deflected coronoid process important in jaw closure) suggestive of the use of gulp feeding as in extant balaenopterids. However, this hypothesis does not consider other anatomical features critical to engulfment feeding such as a nonsynovial, fibrocartilaginous temporomandibular joint and a posteriorly directed mandibular condyle (Berta et al., 2016). It is also possible that *Horopeta* had a sensory organ much like the chemoreceptors in the symphysis of balaenopterids and perhaps eschrichtiids that facilitate gulp feeding in extant balaenopterids (Pyenson et al., 2012). The rib attachments of *Horopeta* suggest an early stage in gulp feeding in which a more complex rib attachment may restrict the volume of water and food taken in a single gulp compared with extant balaenopterids.

Crown Mysticetes

Although still much debated, numerous recent phylogenies (e.g., Boessenecker at al., 2015a, b, c; Gol'din and Steeman, 2015; Marx and Fordyce, 2015; Marx et al., 2016a) include extinct "cetotheres" (and tranatocetids) as crown mysticetes, in addition to the four clades of extant taxa: Balaenidae (bowhead and right whales), Balaenopteridae (fin whales, or rorquals), Eschrichtiidae (gray whales), and Neobalaenidae (pygmy right whales) (figure 3.44, node 9; figure 3.54). Most recent morphology-based phylogenies recognize a dichotomy between a balaenoid clade (balaenids + neobalaenids; figure 3.44,

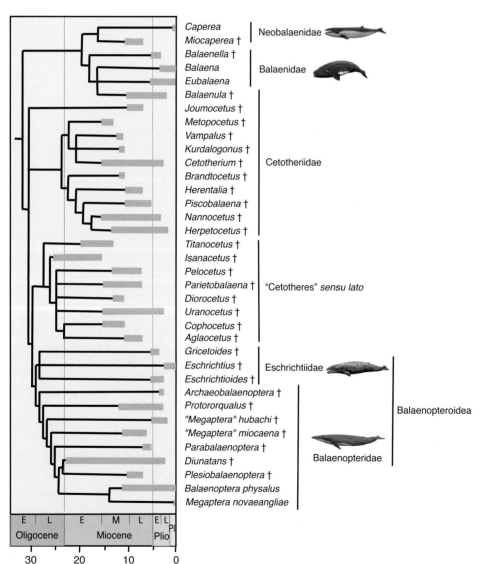

Figure 3.54. Phylogeny of crown mysticetes. From Boessenecker and Fordyce, 2015c.

node 10) and a thalassotherian clade (figure 3.44, node 12) (Boessenecker and Fordyce, 2016). Controversy remains about whether neobalaenids and balaenids are united in a clade, as strongly supported by morphology (e.g., Boessenecker et al., 2015), or whether neobalaenids are sister to all other extant mysticetes, based on robust molecular data. Combined analyses and molecular data sets (e.g., Geisler et al., 2011; Hussain et al., 2012) support another hypothesis: an alliance between Neobalaenidae and Balaenopteroidea in the clade Plicogulae, a name derived from the Latin for "throats with grooves," with reference to grooves on the ventral side of the throat and neck. These grooves are especially well developed in the Balaenopteridae and are used in engulfment filter feeding. The monophyly of Plicogulae contradicts most morphological results, suggesting that this result requires further exploration of data sets.

BALAENIDAE: RIGHT WHALES AND BOWHEAD

The family Balaenidae includes the right whales, *Eubalaena*, and the bowhead whale, *Balaena*. Linnaeus recognized only the genus *Balaena* in his suborder Mysticeti, and early classification systems generally placed all baleen whales into this genus. The living bowhead, *Balaena mysticetus*, occurs in a high-latitude area in the northern hemisphere. Three species of right whales are recognized: the North Atlantic right whale, *Eubalaena glacialis*; the North Pacific right whale, *Eubalaena japonica*; and the South Atlantic right whale, *Eubalaena australis*. Extant right whales have a global distribution (Berta, 2015). Whalers called them the "right" whales to kill because they inhabited coastal waters, were slow swimming, and floated when dead. Balaenids are characterized by large heads that constitute up to one-third of body length. The rostrum is very strongly arched and accommodates extremely long baleen plates (figure 3.44, node 11). Balaenids known as skimmers or intermittent ram-feeders feed by swimming through the water with their mouths open, constantly skimming very small prey, mostly copepods, from the water. The large size of extant balaenids was apparently attained independently in these different clades. Balaenids diverged from other mysticetes approximately 30 million years ago, although the divergence of crown taxa is estimated at 9.82 million years ago (Marx and Fordyce, 2015).

The oldest fossil balaenid, *Morenocetus parvus*, is from the early Miocene (20-18 million years ago) Gaiman Formation of Patagonia, Argentina. The taxon is represented by two subadult specimens and is distinguished by its elongate supraorbital process and triangular occipital shield that extended anteriorly. In addition to this species, a new genus and species of balaenid from the Puerto Madryn Formation is currently under study. *Morenocetus* and the new balaenid are positioned as a basal clade of balaenids, sister to Balaenopteroidea and Cetotheriidae. Stratigraphic calibration of the phylogeny indicates a long temporal gap between *Morenocetus* in the early Miocene and balaenids of the early Miocene and Pliocene (Buono et al., 2014b).

Relatively abundant fossils of later-diverging balaenids are known especially from the late Pliocene of Europe, including *Balaenula astensis* and *Balaena montalionis* (Italy) and *Eubalaena belgica* and *Balaenella brachyrhynus* (Belgium), as well as *Balaenula* sp. from Japan. Described from the North American Pliocene is a nearly complete skeleton of a fossil bowhead, *Balaena ricei*, found in the Yorktown Formation of the eastern United States. Cladistic analysis positions *Balaenula* as sister taxon to right whales in a separate clade from the bowhead. Small size was apparently common among various fossil balaenids, including *Morenocetus* and species of *Balaenula* and *Balaenella*. For example, *Balaenula astensis* has been described as having a length of approximately 5 m (16 ft). Marx and Fordyce (2015) suggest that the disappearance of small balaenids, as well as small balaenopterids and herpetocetines, around 3 million years ago coincided with the onset of glaciation in the northern hemisphere. They further note that changes in glacial cycles would reduce available shelf areas, which was more likely to affect smaller species than larger pelagic taxa.

The genus *Eubalaena* is known from the late Miocene to early Pliocene (6 million years ago) of Japan. The fossil *Eubalaena shinshuensis*, based on a skull and lumbar vertebra, is reported from the Gonda Formation in Nagano Prefecture, Japan. Several as yet undescribed balaenids, including a nearly complete skeleton, are known from the early Pliocene of Japan and the late Pliocene San Diego Formation of California.

NEOBALAENIDAE: PYGMY RIGHT WHALES— BALAENOIDS OR SURVIVING "CETOTHERES"?

One of the most controversial, long-standing debates concerns the phylogenetic position of the small (4 m) pygmy right whale, *Caperea marginata*, found only in the southern hemisphere. The biology of this enigmatic mysticete is poorly known, and most information comes from stranded animals. Most recently, a novel hypothesis proposes that the pygmy right whale might be the last survivor of the extinct Cetotheriidae (Fordyce and Marx, 2013; Marx and Fordyce, 2016), although other recent phylogenetic studies (e.g., Boessenecker and Fordyce, 2015a) ally this taxon with balaenids, supporting a more traditional placement. Additional study using a larger, more comprehensive phylogenetic analysis of mysticetes including both fossil and genomic data is needed to resolve this controversy.

From the time of its initial discovery, *Caperea* has been recognized as having a unique skeletal morphology. It is distinguished from all other mysticetes by a larger, more anteriorly thrusted occipital shield and a shorter, wider, and less arched mouth that accommodates relatively short baleen plates. Other differences distinguishing the pygmy right whale from the balaenids include its having a dorsal fin, longitudinal furrows on the throat (caused by mandibular ridges that might be homologous to throat grooves), coarser baleen, a smaller head relative to body size, a proportionally shorter humerus, and four instead of five digits on the hand.

Prior to 2012, neobalaenids were unknown in the fossil record. In addition to a confirmed new fossil neobalaenid, several possible records now exist. Most interesting, if confirmed as a neobalaenid, is a specimen described by Buono and colleagues (2014a) from Patagonia, Argentina, which would represent the first fossil occurrence of

this lineage in the southwestern Atlantic and the oldest specimen so far reported. The lower jaw specimen collected from the late Miocene (10 million years ago) Puerto Madryn Formation displays a dorsoventrally arched body, a low, bluntly triangular coronoid process, and a posteriorly positioned mandibular foramen similar to that of the extant species *Caperea marginata*. Another questionable record is an ear bone (posterior process only) from the late Miocene (6.2–5.4 million years ago) of Beaumaris, Australia, reported by Fitzgerald (2012b). The first well-documented record of a fossil neobalaenid was described from the late Miocene (8–7 million years ago) Pisco Formation of Peru. This fossil skull was named *Miocaperea pulchra* for its age and similarity to the extant pygmy right whale and for the beautiful preservation of the specimen, which includes baleen (see figure 3.40). *M. pulchra* exhibits characters (long baleen plates and arched rostrum) associated with continuous ram-feeders such as extant balaenids. The occurrence of a fossil neobalaenid 2,000 km north of the northernmost occurrence of the pygmy right whale has been linked to the existence of a coastal upwelling system off the coast of Peru. The stratigraphic distribution and lack of reported neobalaenids from the northern hemisphere is consistent with a southern origin for the *Caperea* lineage, apart from the suggestion of a possible cetotheriid alliance mentioned above.

THALASSOTHERII: CETOTHERIIDAE AND BALAENOPTEROIDEA

Thalassotherii, meaning "sea mammal beasts," is a superfamily-ranked clade comprising two discrete lineages consistently recognized in most recent studies (e.g., Bisconti et al., 2013; Boessenecker and Fordyce, 2015a, b, c): Cetotheriidae (figure 3.44, node 13) and Balaenopteroidea (figure 3.44, node 14). A third, more problematic group consists of stem thalassotherians referred to as "cetotheres."

It is clear that thalassotherians play a critical role in mysticete phylogeny. Although some past studies have placed Cetotheriidae and "cetotheres" outside crown mysticetes, nearly all recent phylogenies place Cetotheriidae and "cetotheres" within crown taxa specifically allied with Balaenopteroidea (e.g., Bisconti, 2015; Boessenecker and Fordyce, 2015a). Most recent studies recognize Cetotheriidae as including *Brandtocetus*, *Cetotherium*, *Herentalia*, *Herpetocetus*, *Joumocetus*, *Kurdalogonus*, *Metopocetus*, *Nannocetus*, *Piscobalaena*, and *Eucetotherium*. Problematic stem thalassotherians are generally recognized as a paraphyletic assemblage, including *Parietobalaena*, *Cophocetus*, *Diorocetus*, *Aglaocetus*, *Isanacetus*, *Thinocetus*, *Halicetus*, *Pelocetus*, and *Titanocetus*. Most of these taxa were described by Remington Kellogg beginning in 1924. The two groups identified by Kellogg are generally distinguished based on the mode of dorsal rostral telescoping. The Cetotheriidae, which include the type genus *Cetotherium*, have rostral bones posteriorly wedged into the frontal in a deep V-shaped pattern, whereas stem thalassotherians have a nearly straight or slightly posteriorly indented junction between the rostral bones and frontal.

Joumocetus shimizui described from the late Miocene (11.6–7.2 million years ago) Haraichi Formation near Yoishi, Japan, is recognized as the geologically oldest and most archaic cetotheriid, as strictly defined (*sensu stricto*). Members of this taxon exhibit a shallower V-shaped wedge between the orbits than other Cetotheriidae, which may represent an intermediate (or primitive) condition. In addition to their occurrence in the Pacific, cetotheriids are known from the Atlantic, including a number of widely distributed Eastern Paratethyan taxa described from the late Miocene of the present-day Black and Caspian Seas. These include *Cetotherium* spp., *Brandtocetus*, *Kurdalagonus*, and possibly *Eucetotherium* (reviewed in Gol'din and Startsev, 2014). A generalized filter-feeding strategy involving different modes of oral suction is suggested for these taxa, based on transversely expanded squamosals, enlarged pterygoids, and thick nuchal crests, indicating a strengthening of temporal and pterygoid muscles.

One of the most speciose cetotheriids is *Herpetocetus*, with at least five small-bodied species known almost exclusively from the Miocene and Pliocene (6.4–2.5 million years ago) of the North Pacific and North Atlantic. The type species, *Herpetocetus scaldiensis*, was named by the Belgian paleontologist Pierre-Joseph van Beneden in 1872, based on a series of unassociated mysticete fossil remains from Antwerp. A new species, *Herpetocetus morrowi*, described from the late Pliocene of southern California by paleontologist Joe El Adli and colleagues (2014), is represented by an ontogenetic series ranging from juvenile to mature adult. This species has a peculiar jaw joint that would have restricted opening of the mouth to 45 degrees, suggesting that it employed a type of oral suction feeding similar to that of the extant gray whale *Eschrichtius robustus*. Other features shared by these taxa and suggestive of lateral suction feeding are possession of short baleen (hypothesized for *Herpetocetus* based on a flattened palate, broad palatal keel, and elevated glenoid fossa) and posterior elongation of the angular process in the lower jaw, increasing the force associated with adduction. *Herpetocetus* species survived at least to the early to middle Pleistocene, making *Herpetocetus* the only extinct genus of mysticetes currently known from the Pleistocene. One of the best-preserved cetotheriid skulls was described by Bisconti (2015) as a new genus and species, *Herentalia nigra*, from the late Miocene of Belgium. His phylogenetic study revealed that *Herentalia* was positioned as a close relative of *Herpetocetus*, although other studies have placed *Piscobalaena* as sister to *Herpetocetus*. Among the more fortuitous discoveries is the first record of fossilized stomach contents for an extinct cetotheriid, found in the late Miocene Pisco Formation of Peru. The record consists of an aggregate of fish remains belonging to the small, schooling clupeiform fish genus *Sardinops* found within the whale's skeleton (figure 3.55).

The genus *Metopocetus*, a cetotheriid, has been positioned in some phylogenies as a transitional taxon linking herpetocetines and other cetotheriines. Several species have been described, including *Metopocetus durinasus* from the late Miocene Calvert or St. Mary's Formation of the eastern United States and a second, morphologically very similar species, *Metopocetus hunteri*, from the late Miocene of the Netherlands. Both species are characterized by development of an unusually large fossa on the ventral surface of the paroccipital process, which has

Figure 3.55. Examples of fish and scales based on micro-CT reconstructions of the fossilized stomach contents of a cetotheriid. Modified from Collareta et al., 2015.

been implicated in articulation of one of the hyoid bones (stylohyal) to the basicranium, which may have functioned in feeding. Cladistic analysis revealed that a third described species, *"Metopocetus" vandelli*, is not closely related to the genus *Metopocetus* (Marx et al., 2016a).

Problematic stem thalassotherians are mostly known from the middle Miocene and Pliocene on both sides of the North Atlantic, although several taxa (e.g., *Isanacetus*, *Parietobalaena*) are also known from the northwestern Pacific (Japan). *Isanacetus* is generally recognized as basally positioned among members of this clade. Some of the largest extinct mysticetes, all from middle Miocene deposits, include *Pelocetus* and *Halicetus* from the northwest Atlantic and *Cophocetus* from the northeast Pacific. One taxon, *Aglaocetus*, had a distribution that extended into the southwest Atlantic. With the exception of *Parietobalaena* and *Aglaocetus*, remaining broadly defined "cetotheres" (*sensu lato*) are known by only a single species. *Parietobalaena* is known by two species, *P. palmeri* and *P. campiniana*, from both sides of the Atlantic—eastern United States (Maryland) and Belgium, respectively—and

from Japan. *Titanocetus sammarinensis* from the middle Miocene of northern Italy was originally described as *Aulocetus* (Bisconti, 2006).

A new family of mysticetes, the Tranatocetidae, positioned as more closely related to eschrichtiids and balaenopterids than the Cetotheriidae, was described by Gol'din and Steeman (2015) but will require a larger taxon and character analysis before confirmation. In addition to *Tranatocetus*, this new family includes *Mixocetus elysius*, *Mesocetus longirostris*, *"Cetotherium" vandelli*, *"Cetotherium" megalophysum*, and *"Aulocetus" latus*. *Tranatocetus argillarius*, previously known as *"Mesocetus" argillarius*, is described from the late Miocene of Denmark based on restudy of a poorly preserved skull as well as new cranial and postcranial material (figure 3.56). Tranatocetids are distinguished from other cetaceans in exhibiting a unique type of telescoping of the rostral bones in which the premaxillae and nasals divide the maxillae on the vertex and the rostral bones are wedged posteriorly on the skull roof, overriding the frontals and interdigitating with the parietals. The postcranial skeleton of *Tranatocetus* cur-

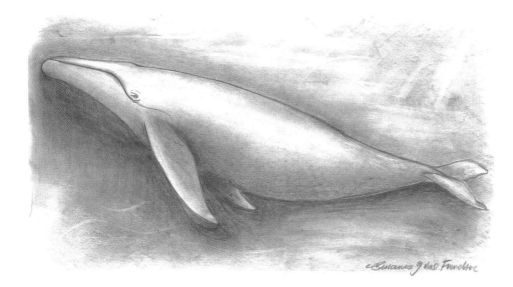

Figure 3.56. Life restoration of *Tranatocetus argillarius*. Illustrated by Susanne Ejdal Ploeg Frandsen. Courtesy of P. Gol'din.

rently under study reveals a distinctively shaped scapula and long deltopectoral crest on the humerus compared with other mysticetes.

Balaenopteroidea: Balaenopteridae (Rorquals) and Eschrichtiidae (Gray Whales). Morphological, molecular, and combined data sets disagree on the relationships between balaenopterids and gray whales. Some combined and morphology-based analyses (e.g., Deméré et al., 2008) support an alliance between eschrichtiids and balaenopterids (Balaenopteroidea), while molecular and other combined data (e.g., Geisler et al., 2011) nest eschrichtiids within balaenopterids, thus rendering the latter nonmonophyletic. Divergence dating places the split of balaenopteroids from other mysticetes at 28.18 million years ago (Marx and Fordyce, 2015).

The Balaenopteridae, commonly called the rorquals (referring to their throat grooves, from the Norwegian *røyrkval*, meaning "grooved whales"), include fin whales, *Balaenoptera physalus*, and the humpback, *Megaptera novaeangliae*, among others. Rorquals are the most abundant and diverse of the living baleen whales (figure 3.44, node 16). They include eight species ranging from the small minke whale, *B. acutorostrata*, at 9 m (29 ft), to the giant blue whale, *B. musculus*. The blue whale has the distinction of being the largest mammal ever to have lived,

reaching 33 m (108 ft) in length and weighing over 160 tons. In the past decade, a new balaenopterid has been discovered, Omura's whale, *B. omurai*, from the Indo-Pacific—a testament to the fact that although we more often hear about vanishing species, a number of new marine mammal species have also been discovered.

Balaenopterids are characterized by a dorsal fin (unlike gray whales and balaenids) and numerous throat grooves that extend from the chin to the umbilicus. Balaenopterids are fast swimmers that lunge at prey and engulf large amounts of water and prey, mostly krill and small fish. In the balaenopterid strategy of lunge feeding, expansion of the mouth and throat is facilitated by the extensive throat grooves. The lower jaw gapes up to 90 degrees, and the ventral throat pouch is distended, inflating like a parachute, and is lined by a slack tongue. With the mouth closed, water is forced out through the baleen so that food is trapped and then swallowed.

The fossil record of the group begins in the middle Miocene, and fossils are reported from North and South America, Europe, Asia, and Australia. The phylogenetically most basal balaenopterid is currently recognized as the early Pliocene (ca. 5 million years ago) *Fragilicetus velponi* from the North Sea, described by Bisconti and Bosselaers (2016). Skull features in this taxon—such as the wide, flat, abruptly depressed supraorbital process of the frontal and the anteriorly constricted supraoccipital, with functional implications related to the anterior position of the temporalis muscle—are interpreted as steps involved in assembly of the specialized morphological features involved in gulp feeding. A fairly well-preserved fossil mysticete skull and skeleton from the middle to late Pliocene of Mount Pulganasco in northern Italy was originally named *"Plesiocetus" cuvieri*. Several later studies variously referred to this specimen as *Cetotherium* and *Cetotherium (Cetotheriophanes)*. Unfortunately, the specimen was destroyed during World War II. A phylogenetic reanalysis based on published illustrations led Bisconti (2005) to recognize this specimen as a new genus and species of balaenopterid, *Protororqualus cuvieri*. In a later publication, Bisconti (2007) named another fossil rorqual from the early Pliocene of northern Italy, *Archaeobalaenoptera castriarquati*, and based on phylogenetic analysis, he suggested that it was the most basal balaenopterid. Its feeding-related structures (e.g., a squamosal with a nearly flat glenoid fossa, as well as a straight lower jaw) indicate that it was not capable of the intermittent ram or lunge feeding characteristic of living members of this group.

There are very few well-documented fossil species of *Balaenoptera*. The oldest known record is a partial immature skull from the uppermost part of the San Gregorio section of the Purisima Formation (early to late Pliocene), described by Boessenecker in 2013 as a new species, *Balaenoptera bertae* (figure 3.57). Phylogenetic analysis suggests that this taxon is most closely related to extant species in a polytomy that includes *B. physalus*, *B. edeni*, *B. acutorostrata*, and *B. bonaerensis*. Another new species of *Balaenoptera*, represented by an ontogenetic series from the San Diego Formation of southern California, is currently under study (figure 3.58). Undescribed balaenopterid material has been recovered from the Pisco Formation in Peru (figure 3.59). Less clear is the origin of the genus *Megaptera*, or even whether it is distinct from *Balaenoptera* spp. In phylogenetic analyses (e.g., Marx and Fordyce, 2015; Marx and Kohno, 2016), two purported fossil species, *"Megaptera" hubachi* from the early Pliocene of Chile and *"Megaptera" miocaena* from the late Mio-

cene of California, were not found to be closely related to the living humpback whale, *M. novaeangliae*. A new late Miocene rorqual based on exceptionally preserved specimens (including traces of mineralized baleen) from the Pisco Formation of Peru was named *Incakujira anillodelfuego*. A total evidence analysis suggests that this new balaenopterid is positioned as sister taxon to the humpback, although this result is not well supported (Marx and Kohno, 2016). The feeding apparatus of *Incakujira* differs from that of modern balaenopterids in possessing restricted Ω (lateral) rotation of the lower jaws that would have limited expansion of the oral cavity. This feature, together with a relatively high baleen plate density, suggested to scientists that *Incakujira* may have been a skim-feeder that targeted relatively small prey.

The family Eschrichtiidae is represented by one extant species, the gray whale, *Eschrichtius robustus* (figure 3.44, node 15). The gray whale is now found only in the North Pacific. A North Atlantic population became extinct in historic time (seventeenth or early eighteenth century). The former distribution of the gray whale in the North Atlantic Basin has been substantiated by Holocene records and by a late Pleistocene fossil occurrence on the seafloor off the coast of Georgia, United States. Collectively, these discoveries at temperate latitudes provide some support for the hypothesis that Atlantic gray whales used a southerly breeding area at the end of a long migratory pathway, just as modern Pacific gray whales use breeding lagoons of Baja California. There are currently two North Pacific subpopulations. The western North Pacific population migrates along the eastern coast of Asia and is extremely rare. The much larger eastern North Pacific population was severely overexploited in the late nineteenth and early twentieth centuries but has recovered sufficiently to be removed from the list of endangered species.

Gray whales are mainly bottom-feeders, employing lateral suction feeding in which the animal rolls to one side (usually the right) and uses its tongue to draw water and prey-laden sediment into its mouth. The short, coarse baleen of gray whales is used to filter bottom-

Figure 3.57. Life restoration of *Balaenoptera bertae*. Illustrated by R. Boessenecker.

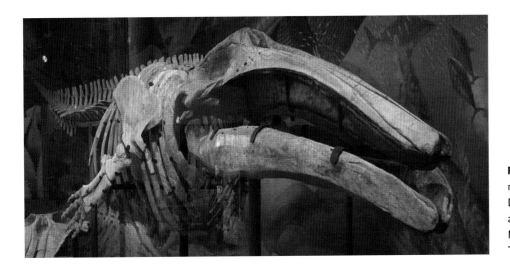

Figure 3.58. Fossil balae-nopteroid skull from the San Diego Formation on exhibit at San Diego Natural History Museum.Courtesy of T. Deméré.

dwelling invertebrates, mostly amphipod and mysid crustaceans.

The earliest known representative of the family is *Archaeschrichtius ruggieroi*, based on a fragmentary lower jaw reported from the late Miocene of Italy. Late Pliocene eschrichtiids are known from rocks in the North Pacific (California and Japan) and the Atlantic (North Carolina); the latter record is based on the fossil *Gricetoides auro-* rae. The most completely described eschrichtiid to date is *Eschrichtioides gastaldii* (previously assigned to *Balaenoptera gastaldi*) from the early Pliocene of northwest Italy (5.3-3.0 million years ago), found to be closely related to the extant gray whale. The holotype is based on a skull and jaws and postcranial elements (vertebrae, ribs, humerus, ulna, metacarpals, and phalanges). The feeding adaptations of *E. gastaldii* as derived from its mandibu-

Figure 3.59. Giovanni Bianucci preparing a balaenopteroid skull at Cerro Colorado, Peru.

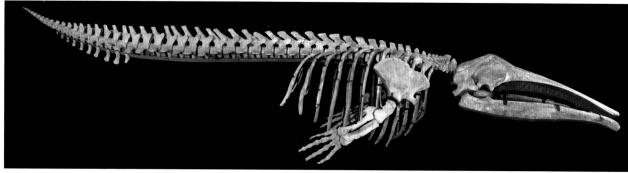

Figure 3.60. Composite fossil eschrichtiid skeleton from the San Diego Formation, California, held in the San Diego Natural History Museum. Courtesy of T. Deméré.

lar morphology—the presence of a postcoronid crest and fossa indicating well-developed adductor muscles such as the temporalis—are similar to those of *E. robustus*, suggesting that it fed using benthic suction similar to that of the living gray whale. The oldest fossil record of the genus *Eschrichtius* sp. is late Pliocene from the western North Pacific (Japan). The skull preserves features commonly associated with *E. robustus* (e.g., a short, steep oc-

cipital shield with distinct occipital tuberosities and robust, posteriorly directed paroccipital processes). A new fossil taxon closely related to the gray whale, represented by excellent, well-preserved skulls and skeletons of a growth series that includes adults and juveniles from the late Pliocene San Diego Formation of southern California, is currently under study (figure 3.60).

4 Aquatic Carnivorans

Pinnipeds and a Bear-like Carnivoran

Pinnipeds, or "fin footed" carnivorans, are widely recognized as monophyletic and include three groups: seals (Phocidae), fur seals and sea lions (Otariidae), and walruses (Odobenidae). The evolution and diversification of pinnipeds involved a number of adaptations, variously developed among the three lineages, including large body size, sexual dimorphism, and deep diving. Pinnipeds originated in the cool waters of the North Pacific during the mid-late Oligocene (30-23 million years ago) (figure 4.1) and diversified worldwide to occupy a wide range of habitats from freshwater lakes to deep oceans. Key cranial features that define pinnipeds include a large opening below the eye socket (infraorbital foramen), suggesting enhanced sensitivity of the snout, and the geometry of several facial bones—including a maxilla that makes a significant contribution to the orbital wall, and a lacrimal that does not contact the jugal (Berta and Wyss, 1994). Among pinniped limb characters are a robust humerus (upper arm) and elongation of the first digit of the hand and the side toes of the foot, modifications that give the fore- and hind-limb bones their distinctive paddle shape, which provided greater swimming power.

Among well-documented fossil pinniped localities of the early to middle Miocene is the Calvert Formation of Maryland and Virginia, which has produced the phocine *Leptophoca* and monachine *Monotherium*. The middle Miocene Sharktooth Hill bone bed of California has yielded otariid and odobenid pinnipeds, as well as desmatophocids. Middle Miocene marine assemblages in the Paratethyan region of central Europe and western Asia (Austria to Kazakhstan) include stem phocines.

The Miocene-Pliocene Pisco Formation of the southern coast of Peru has produced well-preserved skeletons of pinnipeds, including at least 10 species of phocid and otariid seals. The Miocene-Pliocene site of Langebaanweg near Cape Town, South Africa, is the source of a varied terrestrial and marine fauna, including a rich assemblage of remains of the phocid seal *Homiphoca* and as yet undescribed related seals (figure 4.1). The late Pliocene and early Pleistocene San Diego Formation of southern California has yielded a diverse assemblage of marine mammals, including otariids (*Callorhinus*) and odobenids

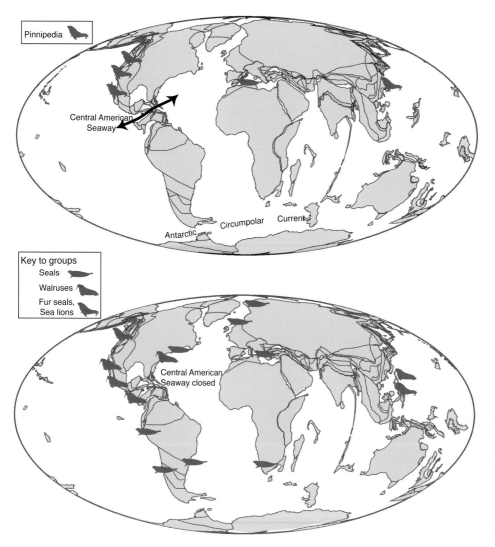

Figure 4.1. Major pinniped fossil localities during the late Oligocene to early Miocene (*top*) and middle Miocene to Pliocene (*bottom*). Modified from Fordyce, 2009. Base tectonic map from www.odsn .de/odsn/about.html.

(*Dusignathus, Valenictus*), as well as mysticetes (*Herpetocetus*, balaenopterids, eschrichtiids, and balaenids), odontocetes (delphinids and phocoenids), and a sirenian (*Hydrodamalis*).

Pinniped Evolution
Body Size

Examination of a large set of data on the body size of extant and fossil pinnipedimorphs revealed a positive relationship between size and age—that is, the increase in pinniped body size was associated with an increase in pinniped generic diversity that began in the middle Miocene (Langhian-Serravallian) (figure 4.2). These changes in pinniped diversity and body size occurred at the same time as or slightly earlier than the rapid increase in cetacean diversity, and both were probably driven by changes in global primary productivity. Changes in pinniped body size were the result of passive selection into vacant niche space rather than active selection for increasing size, since ancestor-descendant comparisons within different lineages have failed to show positive selection for any pinniped clade (Churchill et al., 2015).

Fur and Blubber for Life in the Water

In association with their aquatic living, pinnipeds use fur for insulation. The presence of fur in the earliest-diverging crown otariid, *Callorhinus ursinus*, suggests that dense fur was the ancestral condition and that sea lions

The Basic Anatomy of Pinnipeds

Paleontologists Katrina Jones, Anjali Goswami, and colleagues (Jones and Goswami, 2010; Jones et al., 2015) have shown that the pinniped skull exhibits greater variation in shape (disparity) than the skulls of terrestrial carnivorans, despite pinnipeds' more recent evolutionary origin. Among the pinnipeds, phocids display a much greater diversity of skull morphology than otariids, a reflection of their ecological and reproductive diversity. For example, the bearded seal, *Erignathus barbatus*, with its deep, arched palate and fused lower jaws, is specialized for suction feeding. Males of northern and southern elephant seals, *Mirounga angustirostris* and *Mirounga leonina*, have a large nose that is associated with male fighting and harems. Similarly, male hooded seals, *Cystophora cristata*, have a specialized nasal cavity that forms an inflatable sac when filled with air and is used in courtship display to attract females. The most diagnostic feature of the extant walrus, *Odobenus rosmarus*, is the pair of elongate, ever-growing upper canine tusks found in adult males and females. Elongate tusks evolved in a single walrus lineage, and other fossil walruses (except dusignathines) are tuskless. Fossil pinnipeds also exhibit other unusual morphology, including the fossil monachine seal *Hadrokirus* with its robust jaws and teeth, suggesting a durophagous diet, and the toothless (except for canines) fossil walrus *Valenictus*, a suction-feeding specialist.

The skeleton of pinnipeds reflects adaptations for swimming and terrestrial locomotion in the different lineages. Unlike cetaceans or sirenians, pinnipeds possess both fore and hind flippers. Otariids primarily use the fore flippers and well-developed musculature to increase thrust during propulsion. For example, the supraspinous fossa of the scapula is enlarged, which is correlated with a well-developed supraspinatus muscle. Relatively large neck vertebrae facilitate the extensive neck and head movements of otariids, as shown in the figure. Walruses and phocids use the hind limbs in propulsion and the forelimbs for steering. Their lower back (lumbar) vertebrae have long transverse processes that provide a greater surface area for muscles that move the posterior end of the body. Walruses also use their fore flippers when swimming at low speeds.

Pinniped fore- and hind-limb bones are short and flattened relative to those of terrestrial carnivorans. The limbs are partly enclosed within the body outline to the level of the elbow and ankle. The fingers and toes, or digits, of most pinnipeds are elongated by the development of extensions of cartilage at the end of each digit. On the fore flipper, digit I (thumb) is elongate, whereas on the hind flipper, digits I and V are lengthened. In phocids the hind-flipper digits are more robust, associated with hind-limb swimming, during which these digits act as leading edges of the flipper during the power stroke. Phocids are incapable of turning their hind limbs forward because a tendon passing over the astragalar process of the ankle prevents the foot from being dorsiflexed. Movement of phocids on land is accomplished by undulations of the body, whereas otariids and the walrus are able to turn their hind flippers forward and walk on land. The tail vertebrae in pinnipeds are small and are not used in locomotion.

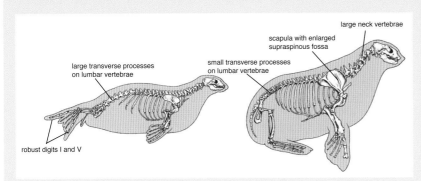

large neck vertebrae

scapula with enlarged supraspinous fossa

small transverse processes on lumbar vertebrae

large transverse processes on lumbar vertebrae

robust digits I and V

Basic anatomy of a pinniped, otariid (*right*) and phocid (*left*).

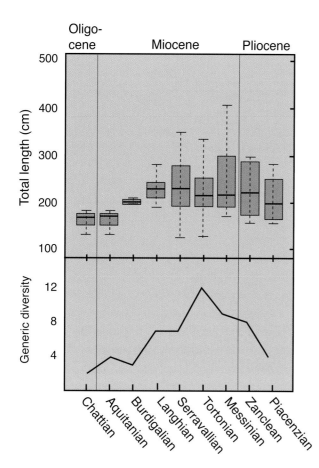

Figure 4.2. Box plots of total length data and generic diversity for pinnipeds through time. From Churchill et al., 2015, fig 7.

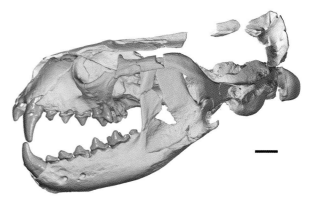

Figure 4.3. Skull and jaw of *Puijila darwini*. Scale bar = 1 cm. From Rybczynski et al., 2009.

secondarily lost fur and developed thick layers of blubber (Liwanag et al., 2012a, b). This most likely occurred several times in the evolutionary history of otariids, given fur seal and sea lion paraphyly. This ancestral state reconstruction of fur is also supported by reconstruction of the paleoenvironment (i.e., cold California currents active during the Miocene and Pliocene) in the North Pacific where fossil members of the *Callorhinus* lineage have been recovered.

Early Beginnings: Pinnipedimorph or Stem Arctoid?

The early stages of evolution are less clear for pinnipeds than for cetaceans and sirenians. A problematic carnivoran fossil named *Puijila darwini* was found in 24- to 20-million-year-old rocks in a lake deposit on Devon Island, Canada. *Puijila* was described by Canadian pale-

ontologist Natalia Rybczynski and colleagues (2009) as a morphological intermediate in the land-to-sea transition of pinnipeds (figures 4.3, 4.4). *Puijila* was just over one meter in length and has been characterized as having the head of a seal and the elongate, streamlined body of an otter. The large, robust canines and jaws suggest *Puijila* had a strong bite and hunted prey both on land and in the water (figure 4.3). Unlike pinnipeds it did not have flippers, and it more closely resembled otters in its limb proportions and possession of a long tail and large, probably webbed, feet. Whether *Puijila* represents a stem pinnipedimorph or a stem arctoid more distantly related to pinnipeds requires further study. Perhaps of greatest significance is that if *Puijila* is confirmed as a pinnipedimorph, this would suggest the Arctic as an early area for pinniped diversification. In terms of habitat, the Arctic was much warmer during the Miocene than now, and the fossil plant record for this time suggests a humid, cool, temperate coastal climate.

Stem Pinnipedimorphs: Sexual Dimorphism and Polygyny?

The evidence indicates that stem pinnipedimorphs such as *Enaliarctos*, *Pteronarctos*, and *Pacificotaria* diversified in shallow bays or inland seas but remained restricted to the eastern North Pacific. The earliest well-represented pinnipedimorph is *Enaliarctos*, known by five species from the early Miocene in the eastern North Pacific. *Enaliarctos*, or "sea bear," was described as a new genus of pin-

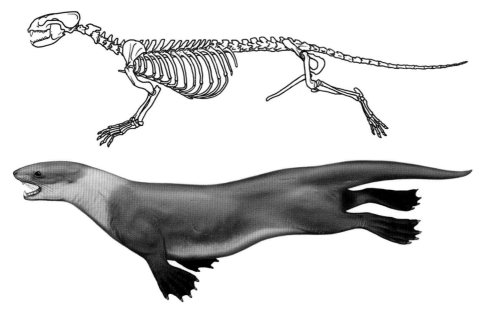

Figure 4.4. Skeletal and life reconstruction of *Puijila darwini*. Illustrated by C. Buell.

niped collected from the Pyramid Hill area near Bakersfield, California, in 1973 by paleontologists Ed Mitchell and Richard Tedford. The Pinnipedimorpha clade originated in the eastern North Pacific (Oregon) during the mid-late Oligocene (30.6-23 million years ago) (figure 4.1). The ancestral dentition of pinnipedimorphs is heterodont, with large, blade-like cusps on the teeth, well adapted for shearing (figure 4.5). Reconstruction of the feeding ecology of *Enaliarctos emlongi* indicates it was a pierce-feeding predator with a diet consisting mostly of fish.

Later-diverging species of the genus *Enaliarctos* show a trend toward a decreasing shearing function of the cheek teeth. These dental trends heralded the development of simple peg-like, or homodont, cheek teeth, which are characteristic of most living pinnipeds. Specific dental features associated with tooth simplification in *Enaliarctos* and not found in their arctoid relatives include reduction in size of the upper molars and cingula, reduction and modification of the entoconid (cusp) on the lower first molar to a crest, reduction in upper postcanine crown size, and a trend toward increasing homodonty (see Churchill and Clementz, 2015). The latest record of *Enaliarctos* is from 20- to 19-million-year-old rocks located along the Oregon coast.

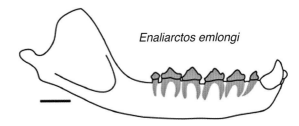

Enaliarctos emlongi

Figure 4.5. Lower jaw of *Enaliarctos emlongi* illustrating tooth morphology and root condition. Scale bar = 20 mm. From Boessenecker, 2011.

Enaliarctos mealsi is represented by a nearly complete skeleton from central California. The entire animal is estimated to have been 1.4-1.5 m (4.6-4.9 ft) in length and between 73 and 88 kg (160-194 lb) in weight, roughly the size and weight of a small male harbor seal. *E. mealsi* was capable of considerable lateral and vertical movement of its vertebral column. In addition, both its fore and hind limbs were modified as flippers and used in aquatic locomotion (figure 4.6). Several features of the hind limb suggest that *E. mealsi* was also capable of maneuvering well on land and probably spent more time near or on shore than do extant pinnipeds. The hinge joints on the hands and feet of *Enaliarctos mealsi* provide some evidence of the use of the limbs to grasp prey and further suggest that prey may have been brought to land for processing

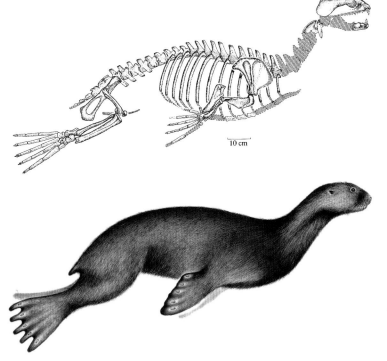

10 cm

Figure 4.6. Skeletal and life reconstruction of the pinnipedimorph *Enaliarctos mealsi.* Total estimated length, snout to tail, is 1.4–1.5 m. *Shaded areas* in skeleton are hypothesized reconstructions of unpreserved bones. From Berta and Ray, 1990.

(Hocking et al. 2017). In a study of skeletal proportions in pinnipeds, Ryan Bebej (2009) hypothesized that *Enaliarctos* was a hind limb–dominated swimmer, since its skeletal proportions are similar to those of hind-swimming phocids.

A later-diverging fossil lineage, more closely allied with pinnipeds than with *Enaliarctos*, includes *Pteronarctos* and *Pacificotaria* from the early to middle Miocene, 19- to 15-million-year-old rocks of coastal Oregon (figure 4.1). A striking osteological feature in all pinnipeds is the geometry of the bones that compose the eye region. In *Pteronarctos*, the first evidence is seen of the uniquely developed maxilla (upper jaw), which contributes to the bones of the eye region. In addition, in *Pteronarctos*, the lacrimal bone and tissues involved in the production of tears are greatly reduced or absent, as they are in pinnipeds. A shallow pit on the palate between the last premolar and the first molar, seen in *Pteronarctos* and later pinnipeds, indicates a reduced shearing capability of the teeth, which together with cusp simplification marks the beginning of a trend toward homodonty (figure 4.7). As is the case for all mammals, tooth simplification in pin-

nipeds appears to be related to the loss in regional expression of the gene *Bmp4*.

Size differences in male and female skulls of *Enaliarctos* suggest that pinnipedimorphs were sexually dimorphic. Given the close correlation between mating system and sex-related skull differences in pinnipedimorphs, such as the expansion of the basicranium and rostrum and the length of the palate relative to the basicranium, Cullen and colleagues (2014) suggest that these ancestral taxa had a harem-based polygynous mating system driven by sexual selection—females selecting larger males with more elaborate musculature for territorial display and male-male competition. These researchers further suggest that climate changes at the time of divergence of *Enaliarctos* from the crown pinnipeds (late Oligocene and early Miocene, 30 million years ago) may have pressured species to develop a polygynous mating system, since harem colonies were likely to be located at upwelling sites along coastal margins that would have concentrated nutrients. There is some support for this idea in modern populations, since Arctic and Antarctic pinnipeds generally do not exhibit sexual dimorphism, and nutrients are more

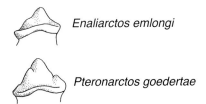   *Enaliarctos emlongi*

Pteronarctos goedertae

Figure 4.7. Comparison of postcanine teeth of *Enaliarctos emlongi* and *Pteronarctos goedertae* illustrating trend toward homodonty. From Boessenecker, 2011.

widely available. There may also be a lesson for the future in this example. If the effects of climate change increase and nutrient levels decrease due to oceanic warming, pressure could be placed on pinnipeds in polar regions to form colonies and develop harem behavior.

Crown Pinnipeds: Phylogeny, Locomotion, and Aquatic Feeding

Given alternative hypotheses for pinniped interrelationships and the relationship of pinnipeds to other carnivorans, elucidating ancestral morphologies in different pinniped groups presents a challenge. Some modification may be required when phylogenetic relationships are settled. A case in point is reconstruction of the evolution of pinniped locomotion. If the sister group to pinnipeds is ursids, which are forelimb swimmers, then the ancestral swimming mode of pinnipeds was forelimb swimming. A musteloid-pinniped alliance would suggest that quadrupedal or pelvic paddling may have been ancestral for pinnipeds, rather than pectoral locomotion. The conflict in pinniped interrelationships involves morphological evidence that allies odobenids and phocids as sister taxa, but molecular and combined data that consistently support a sister-group relationship between odobenids and otariids (see chapter 1, figure 1.5). The story becomes more complicated when extinct desmatophocids, a group of pinnipeds endemic to the North Pacific, are considered, given their uncertain placement among other pinnipeds. Inclusion of desmatophocids in the Otarioidea, as well as a close alliance between otariids and odobenids, suggests that hind-limb swimming in odobenids and phocids may have been inherited from the common

pinnipedimorph ancestor *Enaliarctos*. Alternatively, if desmatophocids, phocids, and odobenids form a clade (Phocomorpha), then hind-limb swimming evolved independently at least twice: once in the enaliarctine lineage and a second time at the base of the Phocomorpha (Desmatophocidae, Phocidae, Odobenidae). This interpretation requires reversals to forelimb swimming in both desmatophocids and dusignathine walruses.

Ancestral state reconstructions indicate that adaptation to aquatic feeding in pinnipeds, evolving from their terrestrial ancestors, involved changes in diverse sensory systems specialized for locating prey, including vision. Extant pinnipeds are among the deepest-diving marine mammals. For example, northern and southern elephant seals, *Mirounga* spp., hold the record for pinnipeds: an ability to dive for more than one hour at a depth exceeding 1,600 m (1 mi) on a single breath of air. A key morphological adaptation in pinnipeds is the size of the bony orbit, a proxy for eyeball size, and this is found to be a predictor of diving depth. Results of a study by paleontologists Lauren Debey and Nick Pyenson (2013) showed that pinnipeds have larger bony orbits than their nearest terrestrial relatives and that deep diving evolved multiple times among crown pinnipeds (e.g., in the monachine clade, especially *Mirounga* spp.). Extrapolating from this data set, Debey and Pyenson found that the largest bony orbit sizes belonged to the middle Miocene fossil pinniped *Allodesmus*, which led these authors to predict that members of this taxon, like elephant seals, were deep divers.

The transition to aquatic feeding in pinnipeds involved gradual simplification of teeth, which has been associated with loss of mastication and transition to pierce feeding. Principal components analysis of dental features by Churchill and Clementz (2015) revealed that the earliest known pinnipedimorphs already exhibited the characteristics of pierce feeding at the base of the crown clade. In crown pinnipeds, multiple feeding strategies have evolved for consuming prey underwater. Initial study of skull and mandible characters in pinnipeds by Adam and Berta (2002) provided evidence for four feeding strategies: filter, grip and tear, suction, and pierce feeding. The extant walrus has long been recognized as a specialized suction-feeder, primarily consuming molluscs, although, as discussed later in this chapter, some fossil walruses were more likely generalist fish-eaters. Phocids were found to employ all four strategies, including a dominant generalist pierce-feeding strategy, characterized by pointed, cusped teeth—a result that was confirmed in a more rigorous quantitative analysis of cranial morphological features (Kienle and Berta, 2015). This same study identified grip and tear feeding by the leopard seal, *Hydrurga leptonyx*, which is characterized by long jaws with enlarged coronoid processes for increased muscle attachment areas, providing a strong bite force. Filter feeding, exemplified by the crabeater seal, *Lobodon carcinophaga*, features highly specialized, multicusped, interdigitating teeth that function as a sieve to filter prey from the water (figure 4.8). Most otariids are also found to be generalist pierce-feeders; presumably, both seal lineages have retained this condition from the common pinnipedimorph ancestor *Enaliarctos*. The next research step requires incorporation of data from fossil pinnipeds and reconstruction of the evolution of feeding strategies in the context of an explicit phylogenetic framework of both extant and extinct pinnipeds.

Otariidae: Fur Seals and Sea Lions
Stem Otariids

The geologically oldest known otariids are *Eotaria crypta* and *Eotaria citrica* from the early middle Miocene (17.1–15 million years ago) Topanga Formation in Mission Viejo, southern California, described by vertebrate paleontologists Robert Boessenecker and Morgan Churchill (2015) and Jorge Vélez-Juarbe (2017) (figure 4.1). *Eotaria* was tiny, with adults only slightly larger than a sea otter. *Eotaria* exhibits extreme tooth simplification and a dentition resembling that of modern otariids. A key feature that distinguishes it from modern otariids is retention of the metaconid on the lower first molar and of a second lower molar. Molecular divergence estimates conflict with the fossil record and place the base of the otariid radiation as much younger, at 8.1 million years ago.

A later-diverging stem otariid, *Pithanotaria starri*, is poorly known by a few fossils from the late Miocene

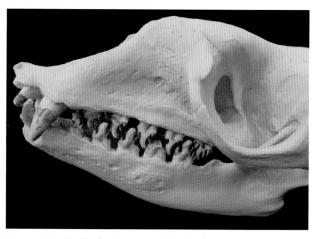

Figure 4.8. Skull of crabeater seal, *Lobodon carcinophaga*, illustrating interdigitating multicusped teeth.

(10–7 million years ago) of California. The holotype is an impression in diatomite of a partial skeleton of an immature individual and preserves limited anatomical information. The same is true for referred material of this taxon. *Pithanotaria* is characterized by double-rooted, simple, crowned cheek teeth, with loss of the second upper molar and metaconid, and a fur seal–like skeleton (figure 4.9). An estimated total length of 126 cm (4.13 ft) makes *Pithanotaria* one of the smallest known otariids.

The best-known stem otariid is *Thalassoleon*, represented by three species from the Miocene of the North Pacific: *T. mexicanus* from Baja California, Mexico, and southern California, *T. macnallyae* from California, and *T. inouei* from Japan (figure 4.1). Features of the skull and teeth indicate that *Thalassoleon* was most likely a fish-eater and used pierce feeding (biting), as do most modern otariids. Also like extant otariids, *Thalassoleon* was a strong forelimb swimmer, but it differed in being more ambulatory on land as judged from its limb proportions (nearly equally proportioned fore and hind limbs), mortised tibia and ankle joint, and lack of modification of foot bones (Deméré and Berta, 2005). Species of *Thalassoleon* are reconstructed as having similar body sizes to *Callorhinus ursinus*.

Figure 4.9. Life restoration of *Pithanotaria starri*. Illustrated by R. Boessenecker.

Crown Otariids

Crown otariids include extant fur seals and sea lions and their close fossil relatives. The traditionally accepted otariid subfamilies, Arctocephalinae (fur seals) and Otariinae (sea lions), are no longer valid. Recent combined morphological and molecular analysis of relationships among otariids by Morgan Churchill and colleagues (2014) supports the northern fur seal (*Callorhinus ursinus*) as the earliest-diverging otariid, followed by a northern sea lion clade (*Zalophus, Eumetopias, Proterozetes*), sister group to a southern clade (*Hydrarctos, Neophoca, Phocarctos, Otaria, Arctocephalus*) (figure 4.10).

Otariids have relatively few characteristic cranial features, including a shelf-like supraorbital process on the skull and simplified dentition with loss of the second upper molar. The earliest-diverging crown otariid, the northern fur seal, *Callorhinus*, has existed as endemic in the North Pacific since the Pliocene. *Callorhinus gilmorei* is known from the middle to late Pliocene, 4- to 2-million-year-old rocks in the North Pacific (Japan and California). There are reports from the late Pliocene Purisima Formation of California of a partial lower jaw referred to *Cal-*

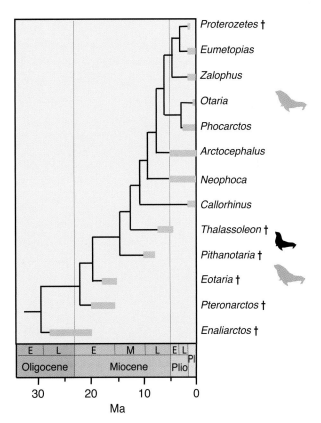

Figure 4.10. Phylogeny of otariids. From Churchill et al., 2014.

lorhinus sp. from the late Pliocene to early Pleistocene (2.1–1.2 million years ago), and dental and postcranial elements referred to *Callorhinus* sp. cf. *C. gilmorei* (Boessenecker, 2011; Boessenecker and Churchill, 2013). These three species (*C. gilmorei*, *C.* sp., and the extant *C. ursinus*) form a morphological continuum, one of the few marine mammal examples of anagenesis, or evolution within a lineage (for another example, see *Thalassocnus* spp. in chapter 6), and exhibit increasingly derived tooth morphology (root condition and size) (figure 4.11). The 4-million-year longevity of this lineage distinguishes it as the longest-surviving pinniped taxon in the Northeast Pacific, occurring along most of the California coastline in addition to Baja California.

The fossil record of otariids is more sparse than the phocid record. *Hydrarctos lomasiensis*, the oldest known southern hemisphere otariid, has been reported from the late Pliocene to early Pleistocene (3.4 million years ago) Pisco Formation of Peru and is positioned as sister taxon to southern fur seals, *Arctocephalus* spp. The fossil sea lion *Proterozetes ulysses*, based on a nearly complete skull, was described from the late Pliocene (2.5 million

years ago) Port Orford Formation near Cape Blanco in southern Oregon. Phylogenetic analysis suggests that it is most closely related to the Northern sea lion, *Eumetopias jubatus*. Description of *Neophoca palatina* is based on a skull reported from the Pleistocene of New Zealand. This fossil species is closely related to the extant Australian sea lion, *Neophoca cinerea*, suggesting that this taxon had a broader distribution in the past and a greater tolerance for colder temperatures (Churchill and Boessenecker, 2016). Other otariid records from the Pleistocene consist of extant genera *Otaria* cf. *O. byronia* from Brazil and Chile and *Arctocephalus* from Brazil and South Africa.

Phocidae: Seals

The most diverse pinnipeds are the phocids, known by 18 extant species. Most morphological and molecular data support two groups of phocids: Monachinae (southern seals, including elephant and monk seals) and Phocinae (northern seals). Both lineages are known from the middle Miocene, approximately 15 million years ago, in the North Atlantic (North America and Europe) although the fossil record may be slightly older (figure 4.1) than

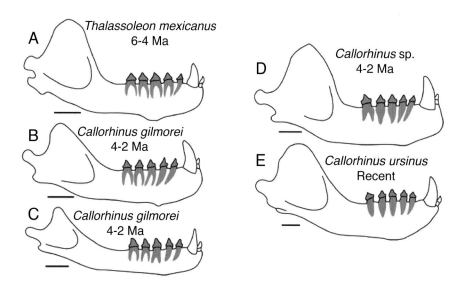

Figure 4.11. Lower jaws (lateral view) and postcanine root morphology in *Thalassoleon* and *Callorhinus* spp. *Callorhinus gilmorei* is from the Rio Dell Formation (*B*), and a holotype specimen is from the San Diego Formation (*C*). *Callorhinus* sp. (*D*). *Callorhinus ursinus* (*E*). The depicted root morphology is based on alveolar morphology, whereas *Callorhinus* sp. is drawn from an X-ray. Missing parts of fossils have been reconstructed. Scale bars = 20 mm. From Boessenecker, 2011. *A*, redrawn from Deméré and Berta, 2005. *C*, redrawn from Berta and Deméré, 1986. *E*, redrawn from Brunner, 2004.

molecular estimates for divergence between monachines and phocines (Higdon et al., 2007). An earlier report of a phocid from the late Oligocene (29-23 million years ago) of South Carolina cannot be confirmed as it is based on very few specimens (two partial femora) with questionable stratigraphic provenance (Koretsky and Sanders, 2002). An important center of late Miocene phocid diversification was the Paratethys, an inland sea formed by the large-scale drying of the Mediterranean Sea, known as the Messinian salinity crisis (figure 4.1) (see also chapter 3).

Phocid seals can be distinguished from otariid seals and odobenids on the basis of several characters of the ear region, including a pachystotic mastoid region and greatly inflated entotympanic. Among postcranial characters is the inability to draw the hind limbs forward under the body due to an ankle joint with a massively developed astragalar process. Relationships among fossil and extant phocids that incorporate both molecular and morphological data are not yet available.

One of the best represented stem phocids is *Devinophoca*, a small-sized seal known by nearly complete skulls and referred postcrania from the early middle Miocene (16.26-14.89 Ma) of Slovakia. However, future work is needed to evaluate the utility of the "ecomorphotype" method employed by vertebrate paleontologist Irina Koretsky and colleagues (e.g. Koretsky and Holec, 2002; Koretsky and Rahmat, 2015) to associate isolated postcrania of this and other Paratethyan Miocene seals to taxa with cranial material based on analogy with extant taxa.

Stem Monachines

The earliest confirmed monachines are known from fossil discoveries in the western North Atlantic. By the late Miocene they had established a circum-Atlantic distribution that included the Mediterranean and Paratethys seas (figure 4.1). Remains of an indeterminate monachine from the middle Miocene of Malta and an even older purported monachine, *Afrophoca libyca*, based on a partial lower jaw from early middle Miocene, 19- to 14-million-year-old rocks in Libya, imply a long history of monachines in the Mediterranean. Other stem monachines include *Pontophoca* from the late Miocene of the North Sea. The Central American (Panamanian) seaway separating North and South America, which was open for most of the Cenozoic (until 11 million years ago, then open again from 6 to 4 million years ago), most likely served as the dispersal route for monachine seals in coastal Peru. This seaway also played a role in the evolution and dispersal of the New World monk seals (*Neomonachus*), as discussed later in this chapter. Also in the late Miocene, an icecap that formed in west Antarctica resulted in major cooling of the Southern Ocean.

South American stem monachine seals include four closely related genera: *Acrophoca*, *Piscophoca*, *Hadrokirus*, and *Australophoca*, and there is some evidence that some of these taxa could be lobodontines. One of the most unusual phocids is *Acrophoca longirostris* from the late Miocene to early Pliocene of Peru and Chile, unique among phocids in having a long, slender, flexible neck and elongate body (figure 4.12). The skull of *Acrophoca* is longer than would be expected relative to its width, a condition known as dolichocephaly. The degree of dolichocephaly in this taxon is greater than in any extant terrestrial carnivorans including canids, implying functional differences in skull shape. The tooth morphology of *Acrophoca* is consistent with generalized pierce feeding, but also includes interdigitating tooth cusps that have been linked with filter feeding. Its diet probably consisted of fish. *Acrophoca* was approximately 1.5 m (nearly 5 ft) in length. An everted pelvis and modified hind flippers suggest that it was a hind limb–dominated swimmer. It was less well adapted for swimming than its lobodontine seal relatives, suggesting that it may have spent more time near shore.

A new monachine seal, *Hadrokirus*, meaning "stout tooth," also from the late Miocene to early Pliocene (5.75 million years ago) Pisco Formation of Peru, was recently described by Amson and Muizon (2014). *Hadrokirus martini* displays sexual dimorphism, as in extant lobodontine seals such as *Hydrurga*. The skull is long and heavily built, and the jaws are robust and stout. The strong masticatory and neck muscles and the robust teeth, some showing breakage and extensive wear (figure 4.13), suggest

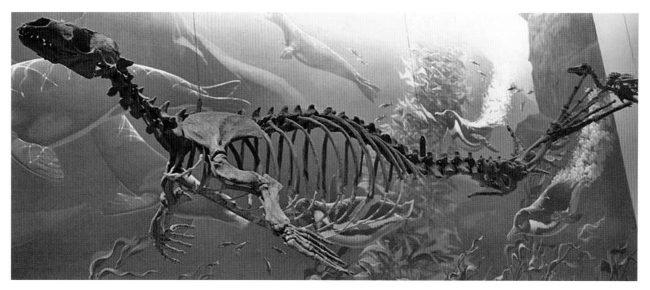

Figure 4.12. Fossil phocid *Acrophoca longirostris* on exhibit at the Smithsonian.

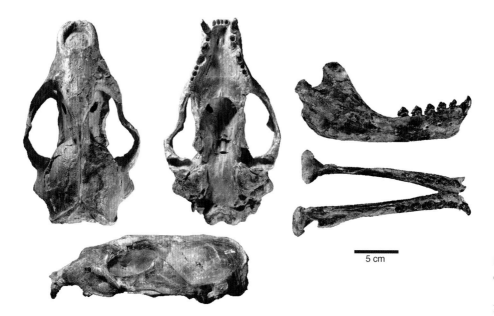

5 cm

Figure 4.13. Skull and mandibles of *Hadrokirus martini*. From Amson and Muizon, 2014.

that the diet of *Hadrokirus* was durophagous and included hard prey items such as molluscs and/or echinoderms. Based on skull measurements, it is likely that *Hadrokirus* approximated the size of an extant Mediterranean monk seal. Powerful neck musculature is suggested by the relatively large atlas and axis vertebrae with broad muscle attachment areas. *Hadrokirus* shows several adaptations (e.g., erect position of the head) that suggest it had greater ability to move on land than extant lobodontine seals and may have spent more time on shore. In terms of phylogenetic position, *Hadrokirus* has been allied with lobodon-

tine seals as the sister taxon of another Pisco Formation seal, *Piscophoca pacifica* (see figure 4.14).

A new dwarf monachine seal, *Australophoca changorum*, was described by Chilean paleontologist Ana Valenzuela-Toro and colleagues (2016), based on diagnostic postcrania from the late Miocene Bahia Inglesa Formation of northern Chile and the Pisco Formation of southern Peru. This seal is smaller than all known fossil and extant monachines, in the size range of the smallest phocines (e.g., Baikal seal, *Pusa sibirica*; Caspian seal, *Pusa caspica*; and ringed seal, *Pusa hispida*).

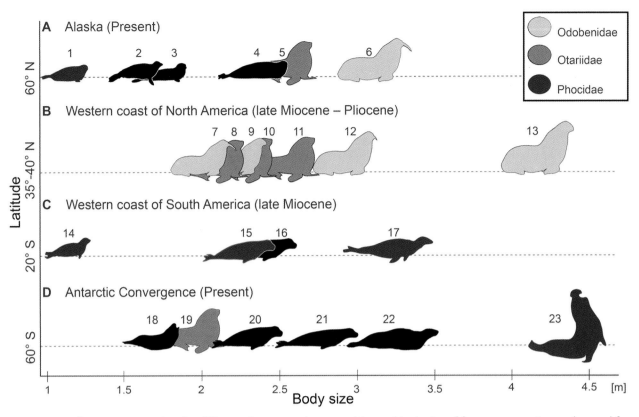

Figure 4.14. Silhouettes comparing the difference in taxonomic composition and body size of four representative modern and fossil communities. 1, *Pusa hispida*; 2, *Histriophoca fasciata*; 3, *Phoca vitulina*; 4, *Erignathus barbatus*; 5, *Eumetopias jubatus*; 6, *Odobenus rosmarus*; 7, *Valenictus chulavistensis*; 8, *Callorhinus gilmorei*; 9, *Dusignathus santacruzensis*; 10, *Pithanotaria starri*; 11, *Thalassoleon mexicanus*; 12, *Gomphotaria pugnax*; 13, *Pontolis magnus*; 14, *Australophoca changorum*; 15, *Hadrokirus martini*; 16, *Piscophoca pacifica*; 17, *Acrophoca longirostris*; 18, *Ommatophoca rossii*; 19, *Arctophoca gazella*; 20, *Lobodon carcinophaga*; 21, *Leptonychotes weddellii*; 22, *Hydrurga leptonyx*; 23, *Mirounga leonina*. From Valenzuela-Toro et al., 2016. Body lengths based on Churchill et al., 2015; Shirihai and Jarrett, 2006. Silhouettes based on illustrations from Reeves et al., 1992; Shirihai and Jarrett, 2006.

The evolutionary history of pinniped communities has recently been examined by Valenzuela-Toro et al. (2016). The fossil record of pinnipeds in South America is dominated by phocids, whereas the pinniped fauna in South America has undergone a turnover and today is dominated by otariids. Valenzuela-Toro and colleagues suggest that this was due to a changing sea level that reduced the number of haul-out sites suitable for phocids and an increase in rocky islands surrounded by a deeper-water environment, which favors otariid seals. Another difference between past and present pinniped communities is the range of body size. Phocid body sizes in the late Miocene assemblage from Chile and Peru are greater than those now seen either in Alaska or in the Antarctic Convergence, although both overall size range and taxo-

nomic diversity for the living pinniped communities are greater (figure 4.14).

Monachine seals from Langebaanweg, South Africa, include *Homiphoca capensis* and as many as six additional species (Govender, 2015). Recent phylogenetic analysis (e.g., Amson and Muizon, 2014) places *Homiphoca* in a clade with South American fossil seals (*Hadrokirus, Acrophoca, Piscophoca*) and lobodontine seals, including crabeater seal (*Lobodon*), leopard seal (*Hydrurga*), and Ross seal (*Ommatophoca*) (figure 4.15). The Langebaanweg deposits and the presence of marine, estuarine, and freshwater taxa indicate a lagoon or estuary that was protected from wave action but still open to the sea. Evidence for *Homiphoca* breeding in the Langebaanweg area comes from the remains of pups and immature animals.

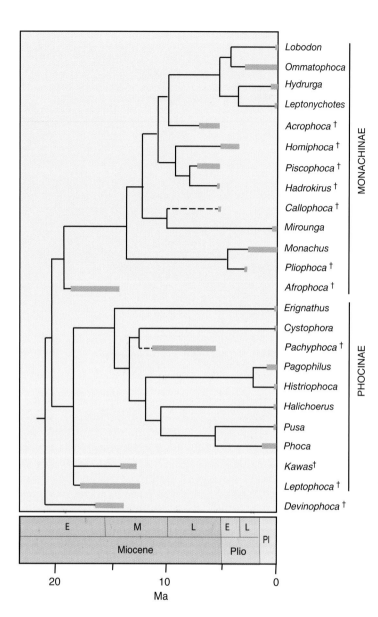

Figure 4.15 Composite phocid phylogeny. Phocinae data from Fulton and Strobeck, 2010. Monachinae data from Amson and Muizon, 2014.

Homiphoca has long nasal bones, voluminous maxilloturbinates, and posterolateral expansion of the maxilla that suggests the presence of countercurrent exchangers, enabling the reduction of heat loss—the latter implying that *Homiphoca* was adapted to living in cold climates. Like the pinniped fauna in South America, the South African pinniped fauna is currently dominated by otariids, in particular the South African fur seal, *Arctocephalus pusillus*. A reduction in island haul-outs may have caused the phocids to move further south where they used ice as haul-out sites, as seen today among Antarctic lobodontine seals.

In addition to the well-documented presence of stem monachines in South Africa and South America, the report of undescribed monachines from the Pliocene of Australia and New Zealand supports a broader distribution of this lineage in the southern hemisphere.

Crown Monachines

Three clades of crown monachines have received consistent, strong support from molecular data: monk seals (*Neomonachus, Monachus*), elephant seals (*Mirounga* spp.), and lobodontine seals (*Hydrurga, Lobodon, Ommatophoca, Leptonychotes*). For the most part, monachines initially were of small body size, smaller than in extant taxa, averaging 180-190 cm (5-6 ft) in length. Later-diverging taxa such as *Monachus*, *Neomonachus*, and *Mirounga* show a trend toward increasing body size. Lobodontines, however, reverse this trend: both increases and decreases in body size occur within this lineage (Churchill et al., 2015).

The best-known Mediterranean fossil monachine is *Pliophoca etrusca*, known by a partial skeleton from the late Pliocene (3.19-2.82 million years ago) of Italy. Phylogenetic analysis by Berta and colleagues (2015) found that *Pliophoca* is closely related to the Mediterranean monk seal, *Monachus monachus*. *Pliophoca* and abundant skeletal remains of another monachine, *Callophoca*, have been reported from the Pliocene (3.4-3.2 million years ago) Lee Creek Mine in North Carolina. Study of morphological and molecular data by Scheel and colleagues (2014) supports a New World genus of monk seals, *Neomonachus*, distinct from *Monachus*. A molecular divergence time for *Neomonachus* spp. was estimated at 3.67 million years ago, which conflicts with the later fossil record reported for the earlier-diverging *Pliophoca*. According to the biogeographic scenario proposed by Scheel et al., the common ancestor of New World monk seals, *Neomonachus*, may have been broadly distributed through the Central American seaway during the late Pliocene. The final closure of the seaway 2.5-2.0 million years ago created allopatric populations of *Neomonachus* in two oceans, giving rise to the Caribbean species, *N. tropicalis*, and the Hawaiian species, *N. schauinslandi*. The fossil record for *Neomonachus* is limited, so we are unable to determine the biogeographic history of this taxon. The earliest fossil record of *N. tropicalis* is from the middle Pleistocene (1.7-1.05 million years ago) of Florida.

Callophoca, at nearly 3 m (9.8 ft), is an exception to the relatively small body size of stem monachines. The sexually dimorphic *Callophoca* has been hypothesized to be involved in the ancestry of elephant seals, but this relationship was recovered with poor support. Other studies have instead placed *Callophoca* within paraphyletic stem monachine taxa (e.g., Berta et al., 2015). Although much is known of the biology of elephant seals, *Mirounga*, they are relatively rare in the fossil record. Previously, a North

American origin of this genus was proposed, with dispersal eastward to Europe, as well to South America through the Panamanian seaway before it closed. However, an improving fossil record of elephant seals from the southern hemisphere suggests a new interpretation of their biogeographic history. Middle to late Pleistocene fossils of *Mirounga* sp. have been reported from Chile, and the oldest record of this lineage is now documented from the early Pleistocene (2.6-2.4 Ma) of New Zealand (Boessenecker and Churchill, 2016).

Lobodontine seals, also known as Antarctic seals, include four monotypic genera: crabeater seal, *Lobodon carcinophaga*; leopard seal, *Hydrurga leptonyx*; Weddell seal, *Leptonychotes weddellii*; and Ross seal, *Ommatophoca rossii* (figure 4.15). This clade diverged from elephant seals approximately 6.9 million years ago and may have dispersed along the western South American coastline. It then spread eastward around Antarctica via the Antarctic Circumpolar Current beginning at least 3.4 million years ago, based on the occurrence of *Homiphoca* in South Africa. Extant genera are not documented in the fossil record, with the exception of *Ommatophoca* reported from the Plio-Pleistocene of New Zealand.

Stem Phocines

The earliest stem phocines have been reported from both sides of the Atlantic: *Leptophoca proxima* (based on systematic re-study of Van Beneden's fossil phocid material by Leonard Dewaele and formerly known as *Prophoca proxima* and *Leptophoca lenis*) from the middle Miocene (16.4-8 million years ago) of Antwerp, Belgium, and the eastern Atlantic (Virginia and Maryland) (Dewaele et al. 2017). This record is slightly older than the molecular estimate for the initial divergence of phocines at 13.0 million years ago. *L. proxima* is a small phocid represented by considerable skeletal material, with an estimated body length of approximately 190 cm (6.2 ft), as reported by Koretsky et al. (2012). These authors described other stem phocines, too, including *Cryptophoca*, *Gryphoca*, *Monachopsis*, *Pachyphoca*, *Phocanella*, *Platyphoca*, *Praepusa*, and *Prophoca*, largely represented by disarticulated, nonassociated postcranial material from various Miocene and Pliocene localities in the region of the Para-tethys. Further systematic study is necessary to confirm the taxonomic status and phylogenetic relationships of these taxa. Koretsky and Ray (2008) also reported the co-occurrence of several of these seals (e.g., *Gryphoca*, *Phocanella*, *Platyphoca*) from the Pliocene Lee Creek Phosphate Mine (Yorktown Formation) of North Carolina. Although rarer and represented only by postcrania, *Gryphoca* and *Platyphoca* are reported from northern Europe, from the late Miocene and early Pliocene North Sea (Denmark, Netherlands). Among the smallest fossil phocines described is *Praepusa boeska* from the Mio-Pliocene of the Antwerp Basin in Belgium. Humeral measurements indicate that this species was smaller than the dwarf fossil monachine *Australophoca*.

Stem phocines are only poorly known from the southern hemisphere. *Kawas benegasorum* was described by Cozzuol (2001) based on an articulated partial skeleton from middle Miocene, 14- to 12-million-year-old rocks of Patagonia, Argentina, although its relationship to other phocines remains unclear. *Kawas* is notable among fossil pinnipeds in being one of only two reported pinnipeds with preserved gut contents, which indicate a diet consisting predominately of bony fish.

Crown Phocines

Among the phocines, consistent strong support is found for recognition of the bearded seal, *Erignathus barbatus*, as a sister group to the remaining taxa, followed by the hooded seal, *Cystophora cristata*. Most recent studies (e.g., Fulton and Strobeck, 2010) find support for the next branch of the tree being the ribbon seal, *Histriophoca*, and harp seal, *Pagophilus*, as sister group to the remaining taxa. For the remaining species, the harbor, ringed, and gray seals (*Phoca*, *Pusa*, and *Halichoerus*), there is disagreement. Several studies position the gray seal as sister taxon of *Pusa*, but in other studies *Halichoerus* clusters within *Phoca* (figure 4.15). The sister-group relationship between the harbor seal, *Phoca vitulina*, and the spotted seal, *Phoca largha*, is consistent and well supported. Evolutionary trends (when compared with monachines) in later-diverging phocines such as the *Histriophoca-*

Pagophilus clade and the *Phoca-Pusa* clade include a reduction in body size (Churchill et al., 2015). This supports, at least in part, the earlier suggestion by paleontologist Andre Wyss (1994) that these clades exhibit juvenilization, their reduced body sizes attributed to heterochrony.

Crown phocines have a fossil record limited to the Pleistocene. *Erignathus* and *Cystophora* have a North Atlantic origin. The earliest record of bearded seal remains is early and middle Pleistocene (ca. 2 million years ago) of Norfolk, England. As has been suggested, the lack of fossils for this lineage may be the result of its early entry into or evolution in Arctic sea ice. The hooded seal is unknown in the fossil record, although the mid-Miocene *Pachyphoca* described from the Paratethyan Basin of Ukraine (Koretsky and Rahmat, 2013) has been included in the cystophorine lineage. *Pachyphoca* is known by two species; the smaller *P. ukranica* is reported to show more adaptations for terrestrial locomotion than its larger relative, *P. chapskii*. The ribbon and harp seals diverged approximately 3.4 million years ago, and their split is hypothesized to have resulted from glacial and interglacial fluctuations that drove allopatric speciation.

Desmatophocidae: An Extinct Lineage

Desmatophocids are an extinct lineage of relatively large pinnipeds discovered in middle Miocene, 23- to 15-million-year-old rocks of the North Pacific. As previously noted, the phylogenetic position of desmatophocids is controversial; they have been allied either with otariids + phocids or with phocids. Two subgroups are recognized: allodesmines and desmatophocines. Although four genera with at least nine species of allodesmines are recognized (*Allodesmus*, *Atopotarus*, *Brachyallodesmus*, *Megagomphos*) from the North Pacific (California, Oregon, Washington, and Baja California) and Japan, a review of the evidence suggests that they are probably oversplit. This is especially true for species of *Allodesmus* from California, three of which are reported from a very limited stratigraphic interval within the middle Miocene Round Mountain Silt (Deméré and Berta, 2002). Desmatophocines are known by a single genus, *Desmatophoca*, with two described species from the Miocene of the North Pacific (Oregon and

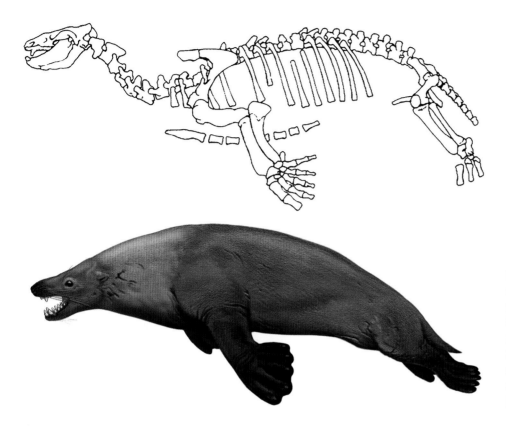

Figure 4.16. Skeletal and life reconstruction of desmatophocid *Allodesmus kernensis* from the Miocene of western North America. Original is 2.2 m (7.2 ft) long (see Berta et al., 2015, for original source and life restoration). Illustrated by C. Buell.

Washington): *D. brachycephala*, found in slightly older rocks and characterized by a wider skull and shorter rostrum, and *D. oregonensis*, with an elongate skull.

Desmatophocids were the first large pinnipeds to evolve. They are characterized by simplified teeth with bulbous crowns, well adapted for eating fish. Desmatophocines had estimated body sizes ranging from the "enaliarctine"-sized *D. brachycephala* (~130 cm [4.2 ft] long) to the sea lion–sized *D. oregonensis* (~200 cm [6.5 ft]) (Churchill et al., 2015). The feeding ecology of desmatophocines was probably similar to that of living otariid seals, and it would have had a generalist diet of fish and squid. Desmatophocines display sexual dimorphism, although to a lesser degree than allodesmines. This provides some evidence of a harem-style reproductive strategy, similar to that of elephant seals, with dominant males controlling females on the breeding beaches.

Allodesmines replaced desmatophocines in the middle Miocene. They are characterized by even larger body sizes. Some reached up to 3 m (9.8 ft) in length, and cranial material from Japan suggests that some individuals attained even larger sizes. Allodesmines are thought to have been large, pelagic predators, ecologically similar to elephant seals. Their large body size possibly rendered them more vulnerable to extinction compared with other lineages such as odobenids, given the latter's greater diversity of body size and dominance during the Neogene.

Large forelimbs and limb proportions provide evidence that the best-known desmatophocid, *Allodesmus kernensis*, was a forelimb-dominated swimmer, although the flexible thoracic and short lumbar regions suggest that a combination of fore- and hind-limb movements was used (figure 4.16). Following from this, it has been suggested that the swimming mode of *A. kernensis* was probably opposite to that of *Odobenus*, with the large fore flippers of *A. kernensis* as the primary propulsive force and its smaller hind flippers and shorter lumbar region secondarily employed in swimming (Bebej, 2009).

Odobenidae: Walruses

The single species of extant walrus, *Odobenus rosmarus*, is distinguished by its elongate upper canine tusks in both

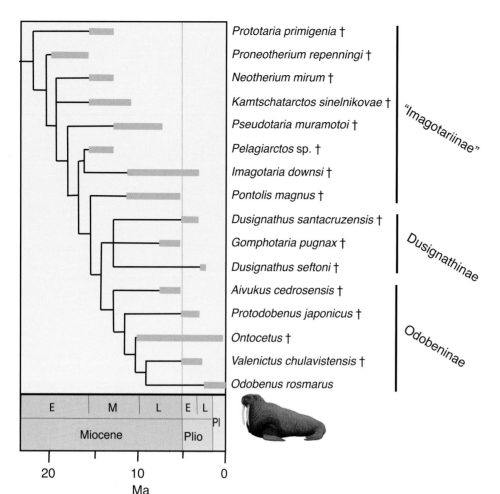

Figure 4.17. Phylogenetic relationships among walruses. From Boessenecker and Churchill, 2013.

males and females. The fossil record of walruses is considerably more diverse, with at least 20 described fossil species in 15 genera. Fossil walruses, unlike extant walruses, show a wide variety of morphological adaptations and body sizes, a reflection of more diverse feeding ecologies and habitats.

Walruses may have evolved earlier but are clearly present by the middle Miocene in the North Pacific, where much of their evolution took place (figure 4.1). In the northern hemisphere, the Bering Strait, a seaway between Alaska and Siberia, first opened as the result of plate tectonic activity during the latest Miocene to earliest Pliocene (5.5–4.8 million years ago). The flow of surface marine water through the strait reversed during the mid-Pliocene (3.6 million years ago), and the resulting south-to-north flow established the modern Arctic ocean circu-

lation pattern. The later opening of the Bering Strait was associated with trans-Arctic biotic interchange, which probably included Atlantic and Arctic molluscs and their chief predators, the odobenine walruses. Upwelling along coastal margins and increased food availability in the North Pacific undoubtedly played a major role in walrus evolution during this time.

Phylogenetic analysis of walruses by Boessenecker and Churchill (2013) and Tanaka and Kohno (2015) supported walrus monophyly and identified three major groups of walruses: stem taxa including "imagotariines," dusignathines, and odobenines (figure 4.17). One of the characters that distinguish walruses from other pinnipeds relates to their unique skull architecture—a large, thick-margined, and elliptical anterior narial opening; another provides increased surface area for attachment of muscles involved in suction feeding—a laterally broad and dorsoventrally thick pterygoid strut.

Stem Walruses: "Imagotariines"

The earliest-diverging stem walruses, with as many as nine genera (*Archaeodobenus, Imagotaria, Kamtschatarctos, Neotherium, Pelagiarctos, Pontolis, Proneotherium, Prototaria,* and *Pseudotaria*), are typically included in the paraphyletic "Imagotariinae." Except for *Archaeodobenus, Pelagiarctos,* and *Pontolis,* most were relatively small, morphologically similar to enaliarctine pinnipeds, and probably not sexually dimorphic.

Prototaria is known by two species, *P. primigenia* and *P. planicephala,* both from the middle Miocene of Japan. The related taxon *Proneotherium repenningi* is known from crania, teeth, and postcrania from the early to middle Miocene Astoria Formation of Oregon. The molariform condition of the premolars of *Proneotherium* represents an early stage in a morphological series extending through *Neotherium* to *Imagotaria* that represents the transition from a shearing to a piercing dentition. *Neotherium mirum,* from the middle Miocene, is well represented by crania and limb material from the Round Mountain Silt in the Sharktooth Hill bone bed. *Neotherium* lacks many of the suction-feeding adaptations of later walruses and has been described as a fish-eating generalist feeder.

Another earlier-diverging imagotariine is *Pseudotaria muramotoi,* known by a skull from the late Miocene (ca. 10-9.5 million years ago) Ichibangawa Formation, northern Japan (Hokkaido). *Pseudotaria* resembles another stem walrus, *Kamtschatarctos sinelnikovae,* described on the basis of a cranial fragment and dentary from the late middle Miocene Kamchatka Peninsula, Russia.

Pelagiarctos from the middle Miocene (16.1-14.5 million years ago) of California is the first large odobenid, estimated to have weighed 350 kg (771 lb), about the size of an adult male South American sea lion. *Pelagiarctos* is also the earliest known pinniped in which fusion of the lower jaws at their anterior ends (symphysis) had evolved; this fusion functioned to reduce strain at the symphyseal joint and strengthen the jaws (figure 4.18). The dentition of *Pelagiarctos* is similar to that of other early-diverging odo-

benids that have well-developed tooth cusps (protoconid, metaconid, and hypoconid) and have been interpreted as generalized fish-eaters also feeding on invertebrates and occasional warm-blooded prey. Support for their generalized foraging habits comes from a study of the enamel ultrastructure of *Pelagiarctos* and otariids, which reveals that they retained the undulating Hunter-Schreger band structure of other extinct and extant carnivorans—differing from the simplified radial or prismless enamel in most cetaceans (Loch et al., 2016). *Pelagiarctos* is the first large odobenid, estimated to have weighed 350 kg (771 lb), about the size of an adult male South American sea lion.

Continuing the trend for large body size among later-diverging "imagotarines" is the fossil walrus *Archaeodobenus akamatsui,* known by more material, including a partial skeleton, described by Tanaka and Kohno (2015) from the late Miocene Ichibangawa Formation of Japan. It is positioned as sister taxon to *Pontolis* and all later-diverging odobenine and dusignathine walruses. *Archaeodobenus* had an estimated body length of 2.8-3.0 m (9-9.8 ft) and a weight of 390-473 kg (860-1,042 lb). The coexistence of two morphologically similar odobenids (*Pseudotaria* and *Archaeodobenus*) in a geographically restricted area may have been the result of a marine regression (fall in sea level) that would have provided reproductive isolation and driven speciation that led to the marked diversification of walruses occurring at this time.

Figure 4.18 Life restoration of the head of *Pelagiarctos* sp. From Boessenecker and Churchill, 2013.

A related, later-diverging taxon, *Pontolis*, was a very large walrus with a skull length nearly twice that of adult males of the modern walrus, *Odobenus rosmarus*. Some stem walruses such as *Pontolis* show considerable diversity in skull shape, which suggests a different evolutionary pattern for walruses than for other pinnipeds. One of the most completely known odobenids is *Imagotaria downsi* from the early Miocene (12–10 million years ago) of California, the sole species in the genus. Conical, unworn teeth and lack of a vaulted palate indicate that *Imagotaria* probably fed on fish. The large sample size and numerous specimens provide evidence of sexual dimorphism in this taxon.

Fish- and Suction-Feeding Walruses: Dusignathines and Odobenines

Dusignathine and odobenine walruses diverged during the late Miocene (figure 4.17). Dusignathine walruses remained endemic to the North Pacific, but odobenines (the modern walrus lineage) underwent dramatic diversification, dispersing from the North Pacific into the North Atlantic via the Central American seaway. According to the latest biogeographic scenario proposed by Kohno and Ray (2008), based on late Pliocene fossil records of odobenines from Japan, this lineage did not become extinct in the North Pacific as earlier thought. Instead, it may have continued to diversify in that region into the Pleistocene, dispersing to the North Atlantic during the late Pliocene.

Tusk development occurs in both dusignathine and odobenine walruses. Dusignathine walruses include species of *Dusignathus* and *Gomphotaria*, which had enlarged upper and lower canines. Within the dusignathines, the tusks of *Dusignathus seftoni* and *Gomphotaria* are not only greatly enlarged but also procumbent. *Dusignathus* is known by two species: *D. santacruzensis* from the Purisima Formation of northern California and *D. seftoni* from the late Pliocene San Diego Formation of southern California. Additional postcrania (cervical vertebra and forelimb elements) referred to *Dusignathus* sp. cf. *D. seftoni* have been reported from the Purisima Formation. *Gomphotaria pugnax* is the most completely known dusignathine. The large skull size (~47 cm [1.5 ft]) indicates

an enormous pinniped, second in size only to *Pontolis* (60 cm [1.9 ft]). Distinctive features include large, procumbent upper lateral incisors, high, elevated sagittal crest, and relatively small eyes. Wear on the tusks and extensive tooth breakage in *Gomphotaria* suggest that this walrus broke shellfish open instead of sucking them out of their shells like the extant walrus. Dusignathine and odobenine walruses possess the best-known and most extreme record of simplification of dentition. During the late Pliocene, *Dusignathus* coexisted with the odobenine *Valenictus*, and it is likely that they divided up the resources by feeding on different prey items. *D. seftoni* had a complete set of teeth and was probably specialized for eating squid and fish.

The genus *Aivukus*, named for *aivuk*, Inuit for "walrus," is known only from a single described species, *A. cedroensis*, from the late Miocene of Baja California. This taxon is positioned as the earliest-diverging odobenine. It is characterized by an elongate, slender rostrum, small lower canines relative to upper canines, peg-like cheek teeth, and unfused mandibular symphysis.

Another early-diverging odobenine is *Protodobenus japonicus*, represented by a well-preserved skull and jaws from the early Pliocene (5.0–4.9 million years ago) of Japan. Both *Aivukus* and *Protodobenus* were tuskless and possibly piscivorous odobenines, as judged by the lack of a highly vaulted palate and wear facets on the anterior teeth.

The most completely known odobenine walrus, *Valenictus chulavistensis*, was described by San Diego vertebrate paleontologist Tom Deméré (1994). It was large-bodied, had a long and deeply arched palate, and was toothless with the exception of the large upper canines. This adaptation may have allowed *Valenictus* to suction-feed on marine invertebrates (especially molluscs) on the seafloor. It has been suggested that incisor loss in odobenines may be related to allowing an oral pathway for suction feeding (figure 4.19). A robust humerus and greatly enlarged medial entepicondyle suggest that *V. chulavistensis* employed a greater degree of forelimb flexion during swimming than the modern walrus, and more detailed study of locomotor adaptations in this taxon are clearly

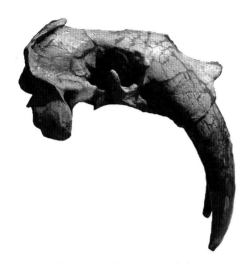

Figure 4.19. Skull and jaws of *Valenictus chulavistensis* on exhibit at the San Diego Museum of Natural History. Skull length is 400 mm (1.3 ft). Courtesy of T. Demére.

warranted. The limb bones of *V. chulavistensis* are osteosclerotic and pachyostotic, important in maintaining horizontal trim in the water.

The closest relative to the *Valenictus* + *Odobenus* clade is the fossil walrus *Ontocetus*. This was initially thought to be a cetacean; the name comes from the Greek *ontos*, meaning "being," and *ketos*, "whale" or "large sea animal." Previously, many scientists attributed Pliocene walruses to several different taxa (*Alachtherium, Trichecodon, Prorosmarus*), which paleontologists Naoki Kohno and Clayton Ray (2008) have since synonymized into the single genus *Ontocetus*. This genus shows extreme sexual dimorphism, which suggests it used a highly polygynous breeding system in land-based rookeries. *Ontocetus* is most commonly represented by tusks that can be identified on the basis of their osteodentine core, which has a globular texture. The outmost portion of the tusk is composed of a very thin layer of cementum. The tusks of this taxon differ from those of other odobenines in having a stronger posterior curvature and grooves along their length, as well as being more compressed laterally. The differing tusk size between members of this taxon from Florida and those from Virginia and North Carolina suggests segregation of males and females during the seasonal feeding migration, which is nearly universal among sexually dimorphic pin-

nipeds. The development of a lesser trochanter on the femur suggests a high mobility on land, as in otariids.

Study of Pliocene walruses has improved our understanding of their biogeography. A species of *Ontocetus* appears first in the early Pliocene of the western North Pacific, supporting the suggestion that the genus originated in the North Pacific and dispersed to the North Atlantic via the Central American seaway. *Ontocetus emmonsi* from the Pliocene of the North Atlantic is hypothesized to have dispersed from eastern North America along the North Atlantic just after closure of the Central American seaway. The North Atlantic population of *Ontocetus* disappeared by the end of the late Pliocene or early Pleistocene, before arrival of the extant walrus, *Odobenus rosmarus*, from the North Pacific via the Arctic Ocean during the late Pleistocene (Demére et al., 2003; Kohno and Ray, 2008).

Other Marine Carnivorans: Bear-like Aquatic Carnivoran (*Kolponomos*)

The large extinct carnivoran *Kolponomos* is known by two species: *K. clallamensis* from the Miocene of Washington and *K. newportensis* from coastal Oregon. A maxillary fragment with alveoli for two large teeth, found on Unalaska Island, Alaska, has been referred to cf. *Kolponomos*. Comparisons of this specimen with described species limited to the early Miocene support an early Miocene age for the Unalaska specimen.

As related by the naturalist David Raines Wallace in his book *Neptune's Ark* (2007), the back of the skull and some vertebrae and foot bones of *K. newportensis* were first discovered in a concretion near Newport, Oregon, by the amateur collector Douglas Emlong in 1969. Quite remarkably, eight years later, Emlong located the concretion's other half with the cranium and jaw, and he recognized that the fossil came from *Kolponomos*. At that time *Kolponomos* was thought to be a type of marine raccoon relative described in 1960 by vertebrate paleontologist R. A. Stirton of the University of California Museum of Paleontology, based on a skull and jaws collected from the Miocene Olympic Peninsula of Washington. The relationship of *Kolponomos* to other carnivorans has been problematic. The current best-supported hypothesis places *Kolponomos* as an ursoid, most closely related to members of the extinct paraphyletic family Amphicynodontidae, which includes *Amphicynodon, Pachycynodon, Allocyon,* and *Kolponomos. Kolponomos* and *Allocyon* are hypothesized to be the stem group from which the Pinnipedimorpha arose (Tedford et al., 1994). Shared derived characters that link *Allocyon, Kolponomos,* and the pinnipedimorphs include details of the skull and teeth. However, the position of *Kolponomos* is unclear because it was excluded from a recent analysis of basal arctoids.

Kolponomos had a massive skull with a markedly downturned snout and broad, crushing teeth. It inhabited nearshore marine rocks. The crushing teeth were suited to a diet of hard-shelled marine invertebrates such as crabs or clams, which it obtained by using its powerful neck muscles to pry and twist them off the substrate, then using its anterior teeth to obtain the soft parts, as sea otters do (figure 4.20).

The unique durophagous adaptation of *Kolponomos,* similar to that of marine otters, remained untested until recently. Using multiple lines of evidence including finite element and geometric morphometric analyses of the lower jaw, as well as observations of cheek-tooth occlusal wear, paleontologist Jack Tseng and colleagues (2016) showed that *Kolponomos* and the saber-tooth felid *Smilodon* convergently share mandibular shape and a similar anchoring function of the mandibles. The *Kolponomos* jaw more closely resembles that of the brown bear, *Ursus americanus,* as reflected by unilateral bite sim-

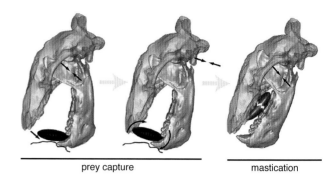

prey capture   mastication

Figure 4.20. Finite element simulations of *Kolponomos* bite force. *Left,* anchor bite (lower canine teeth and the anteriorly buttressed mandible as an anchor). *Center,* neck-assisted torque applied to dislodge substrate-bound hard-shelled prey. *Right,* unilateral crushing bite employing first lower molar. From Tseng et al., 2016.

ulations. Simulation results further revealed that *Kolponomos* crushed prey with an emphasis on jaw stiffness, in contrast to sea otters, which emphasize force. A unique prey capture–mastication sequence was proposed for *Kolponomos* (figure 4.20). Initially, prey capture involved anchoring and wedging of the lower incisors and canines between the prey and the substrate, followed by closing of the mouth, so that the upper and lower anterior teeth bracketed the shell of the prey. Next, high torque was applied by strong neck muscles, with the anterior teeth acting as a fulcrum to dislodge the prey from the substrate. Finally, in crushing bites, the prey shells were crushed with otter-like cheek teeth, but this was achieved with a higher mandibular stiffness and lower mechanical efficiency than estimated in sea otters.

5 Crown Sirenians and Their Desmostylian Relatives

Crown sirenians include two lineages: manatees (trichechids) and dugongids (dugongs). Since the Eocene they have typically inhabited warm, nearshore marine waters, although some species (manatees) frequent or are confined to fresh water, and one lineage (hydrodamalines) is adapted to cool temperate water. Dugongids, the most successful and diverse sirenians, had a worldwide distribution and a long history, especially in the western Atlantic and the Caribbean, but at present they are represented by a single species: the extant *Dugong dugon*, found in the Indo-Pacific region.

The earliest-known dugongids were present in Africa in the middle Eocene. Their record in the Oligocene is poor, possibly the result of colder water temperature. Warmer conditions in the late Oligocene could explain their rise in diversity. In the early Miocene, stem dugongids (halitheriines) occupied the western North Atlantic, with several lineages (e.g., *Metaxytherium*) dispersing into the eastern North Pacific through the Central American (Panamanian) seaway (figure 5.1, top). Fossil evidence of sirenians from the central Indo-Pacific is limited, although the Miocene record of an indeterminate sirenian from a limestone cave provides the earliest evidence of sirenians in Australasia, as well as being the earliest mammal recorded from New Guinea. A cold water–adapted hydrodamaline and a dugongine occupied the eastern Pacific coast at this time. The last of the hydrodamaline lineage, which included Steller's sea cow, *Hydrodamalis gigas*, flourished briefly in the cool waters of the North Pacific during the Plio-Pleistocene, ranging from Baja California, Mexico, to the Aleutians. By the Pleistocene, *Dugong* had entered the Pacific, remaining restricted to the tropical latitudes in the Indo-Pacific.

The evolution of manatees is tied to Caribbean paleobiogeography (figure 5.1, bottom). In the middle Pliocene (4–3 million years ago), the Central American seaway closed, with emergence of the isthmus of Panama cutting the Caribbean-Pacific link. In South America, uplift of the Andes during the later Miocene and early Pliocene (6–4 million years ago) created erosion and runoff of nutrients into river systems, which resulted in an abundance of abrasive grasses. By the middle Miocene, trichechids appeared in South

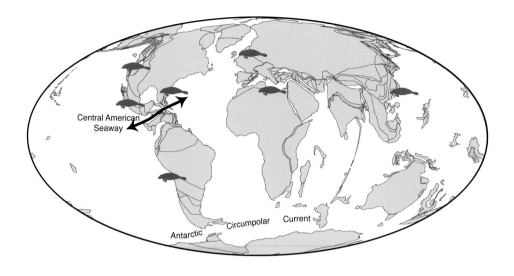

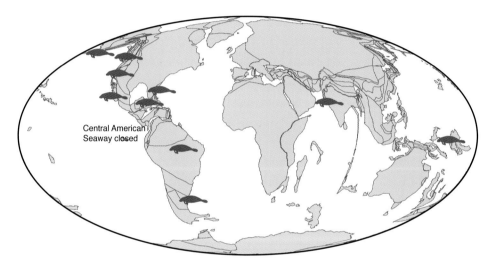

Figure 5.1. Major localities of crown sirenians during the Oligocene and early Miocene (*top*) and middle Miocene to Pliocene (*bottom*). Modified from Fordyce, 2009. Base tectonic map from www.odsn .de/odsn/about.html.

America, restricted to freshwater or estuarine environments (figure 5.1, bottom). They adapted to this new food resource, first through the evolution of thickened tooth enamel, then by a shift to smaller, numerous teeth that were continually replaced. In the late Pliocene or early Pleistocene, trichechids invaded the western Atlantic-Caribbean region, possibly the result of extinction of dugongines in that region in the middle to late Pliocene.

Dugongidae: Dugongs

The paraphyletic family Dugongidae includes the Dugonginae and extinct Hydrodamalinae plus the *Metaxytherium* clade and the paraphyletic extinct "Halitheriinae" (figure 5.2). Ongoing fossil discoveries are revealing a sig-

nificantly greater taxonomic diversity of sirenians than previously known. The earliest dugongids appear in the middle and late Eocene of the Mediterranean; these include *Eotheroides* and the "halitheriines" *Eosiren* and *Prototherium*. The earliest members of the *Metaxytherium* spp. + Hydrodamalinae clade are found in the middle Eocene of the Caribbean and western Atlantic (figure 5.3). In contrast to prorastomids and protosirenids, early dugongids show progressive reduction of the pelvic girdle and hind limbs. These early dugongids, as well as Steller's sea cow and extant dugongs, were marine animals feeding on sea grasses.

The stem dugongid *Eotheroides*, represented by at least six species, is known from the middle Eocene (Lutetian)

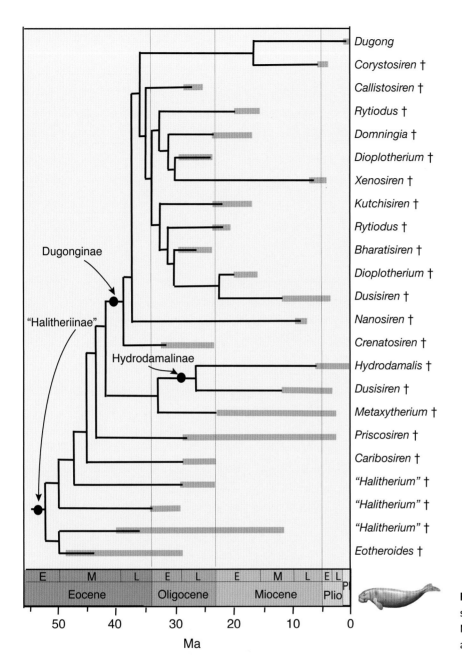

Dugong

Corystosiren †

Callistosiren †

Rytiodus †

Domningia †

Dioplotherium †

Xenosiren †

Kutchisiren †

Rytiodus †

Bharatisiren †

Dioplotherium †

Dusisiren †

Nanosiren †

Crenatosiren †

Hydrodamalis †

Dusisiren †

Metaxytherium †

Priscosiren †

Caribosiren †

"Halitherium" †

"Halitherium" †

"Halitherium" †

Eotheroides †

Dugonginae

"Halitheriinae"

Hydrodamalinae

| E | M | L | E | L | E | M | L | E | L |
| | | | | | | | | | P |

Eocene   Oligocene   Miocene   Plio

50   40   30   20   10   0

Ma

Figure 5.2. Phylogeny of stem and crown Dugongidae. Modified from Vélez-Juarbe and Domning, 2015.

of Egypt, Madagascar, and Pakistan. The oldest known species of *Eotheroides* is *E. aegyptiacum* from the Mokattam Hills near Cairo. *E. babiae* is from Kachchh, India, and *E. lambondrano* is from Ampazony, Madagascar. Another two species, *E. clavigerum* and *E. sandersi*, from Wadi Al Hitan in the western Fayum desert of Egypt, are well represented by skulls, lower jaws, and skeletons with well-developed pectoral limbs and pelvic girdles, as de-

scribed by paleontologists Iyad Zalmout and Phil Gingerich (2012). Specimens of these latter two species are phylogenetically closest to *E. aegyptiacum*. They were medium to large dugongids, ranging in length from 1.5 to 2.5 m (4.9–8.2 ft). Skull bones show a deflected rostrum with very small to intermediate-sized tusks. The pelvis is highly reduced and exhibits morphology that suggests sexual dimorphism. The tail is flat, indicating the pres-

The Basic Anatomy of a Sirenian

Unlike the stem sirenians (amphibious quadrupeds), the crown sirenians possess enlarged premaxillae that form a downturned rostrum and a massive lower jaw with a downturned symphysis. The dugong skull differs from the manatee skull in having a more strongly downturned rostrum (as shown in the figure), an adaptation for feeding on sea grasses on the seafloor. The manatee rostrum and that of *Hydrodamalis* is only slightly deflected, an adaptation for feeding higher in the water column. The massive oil-filled bones in the cheek region, implicated in sound reception, differ from the acoustic fats involved in sound conduction in cetaceans, suggesting that different lipids may be involved in sound conduction in sirenians. The eyes are small, and their lack of well-developed ciliary muscles (needed for close focus) is consistent with their poor near vision.

The external ears are small openings in the sides of the head. The nostrils are located dorsally and at the anterior end of the snout. The most striking feature is the fleshy oral disk, the greatly expanded upper lip between the mouth and nose, which is covered with vibrissae that function to bring food to the mouth. Most sirenians have bunodont/lophodont molars more like those of humans and pigs (an exception is the extant dugong, with peg-like molars), located in the back of the jaws. Chewing occurs by grinding plates in front of the teeth on the roof of the mouth. Steller's sea cow lacked teeth but had grinding plates for feeding on algae.

All sirenians have very sparse, short, fine sensory body hairs. These body hairs are specialized cells associated with nerve cells that are stimulated by water movement. The body hairs seem to be analogous to the lateral line

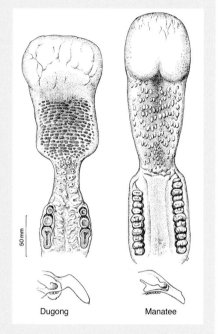

Comparison of grinding plates and snout deflection in a dugong and manatee. From Marsh et al., 1999, with permission from Wiley and Sons, Inc.

systems of fish and some amphibians, which are used to detect underwater objects such as shorelines and sandbars

ence of a fluke. A distinguishing feature of the Malagasy species, *E. lambondrano*, is its small size with a skull length of only 270 mm (10.5 in). Based on recent taxonomic revision, the European genus *Prototherium* is known by two species from Spain and Italy, with new species *P. ausetanum* positioned as sister taxon to *Eotheroides aegyptiacum* (Balaguer and Alba, 2016). More comprehensive phylogenetic analysis is warranted, however, since this same study found Eocene stem dugongids *Prototherium*, *Eotheroides*, and *Eosiren* to be paraphyletic.

The paraphyletic subfamily "Halitheriinae" includes the genera *Halitherium*, *Prototherium*, *Eosiren*, *Caribosiren*, *Metaxytherium*, and *Priscosiren*, known from Europe, North Africa, and the western Atlantic-Caribbean region (areas occupied by the former Tethys Sea). *Eosiren* is

known by three species from the Eocene-Oligocene of Egypt: *E. imenti*, *E. libyca*, and *E. stromeri*. *Eosiren* displays a greater degree of pelvis reduction than *Eotheroides* and *Protosiren*. Both *Eosiren* and *Eotheroides* differ from *Protosiren* in possessing osteosclerotic and pachyostotic ribs and more reduced hind limbs. These stem dugongids probably fed in shallower water than *Protosiren*, based on their more rigid rib cage counterbalanced by bone density and thickening (pachyosteosclerosis).

Priscosiren atlantica from the early Oligocene of the western Atlantic and Caribbean region has been described as close to the ancestor of both the *Metaxytherium* + Hydrodamalinae clade and the Dugonginae clades. Members of this taxon are intermediate in size, approximately 2 m (6.5 ft) in length. They retain a suite of ancestral char-

and the approach of other animals. Neither dugongs nor manatees have a blubber layer, but they possess thick skin and migrate to warmer waters when the temperature drops below 20°C (68°F). Seven neck vertebrae are present except in *Trichechus* spp., which has only six. Sirenians have experienced an elongation of the thoracic vertebrae that is important in trunk stability, which has resulted in shortening of the lumbar region.

Sirenians have modified paddle-like forelimbs, as shown in the figure below. The forelimbs are primarily used in steering. The shoulder, elbow, wrist, and finger joints are freely moveable. The forelimbs of Steller's sea cow are short and fingerless, ending in a blunt and hook-like process, and were probably used to pull the body forward along the sea bottom.

The evolutionary reduction and loss of hind limbs in sirenians parallels that in whales, with extant sirenians having only a vestigial pelvis. Manatees differ from dugongs in having a rounded, paddle-shaped tail (fluke); dugongs have a triangular tail like those of whales. Sirenians, like cetaceans, use the horizontally expanded tail to create propulsion. The skeleton of sirenians typically is both pachyostotic and osteosclerotic, adaptations for providing ballast to maintain neutral buoyancy.

A study of the loss of pelvic fins in three-spined stickleback fish and the observation that sirenians and sticklebacks share asymmetry in their pelvic bones (those on the left side are larger than those on the right) led researchers to suggest that similar genetic mechanism may underlie pelvic reductions in the two groups. Gene expression studies showed that a mutation in the *PitX1* gene was responsible for pelvic reduction in sticklebacks and hind-limb loss in laboratory mice. Although more study is needed to link this mutation with hind-limb loss in manatees and dugongs, the findings support other evidence that such genes have evolved independently and repeatedly in distantly related vertebrate lineages.

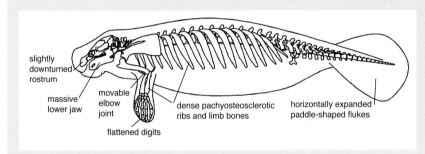

slightly downturned rostrum

massive lower jaw

movable elbow joint

flattened digits

dense pachyosteosclerotic ribs and limb bones

horizontally expanded paddle-shaped flukes

Anatomical features of a sirenian (manatee). Modified from Rommel and Reynolds, 2009.

acters not seen in most other later-diverging dugongids. The best-known "halitheriine" genus, *Metaxytherium*, was widely distributed in both the North Atlantic and Pacific during the Miocene. It is represented by at least eight described species. A west Atlantic-Caribbean origin of the oldest, basal-most described species, *Metaxytherium albifontanum*, has been proposed. A later and more derived species of this taxon was reported from the late Pliocene Mediterranean Basin. European *Metaxytherium* spp. form an ancestral-descendant lineage (*M. krahuletzi, M. medium, M. serresii, M. subapennium*) with little change through the Miocene, but with large tusks evolving during the Pliocene. *Metaxytherium* had a strongly downturned snout and small upper incisor tusks. Most "halitheriines" were bottom-feeding animals that consumed rhizomes (root-like stems) of small to moderate-sized sea grasses as well as sea-grass leaves, according to tooth isotope values (figure 5.4). Some, however, may have fed on C3 plants of freshwater systems.

The extinct Hydrodamalinae group includes the paraphyletic genus *Dusisiren* and is distinguished as the lineage that invaded cool waters, developed a large body size as an adaptation to cool climates, and led to the recently extinct Steller's sea cow, *Hydrodamalis gigas* (figures 5.5, 5.6). These trends are attributable in part to a general cooling of the North Pacific and the replacement (to a large extent) of sea grasses by kelps. The earliest hydrodamaline, *Dusisiren reinhardi*, appeared in the early to middle Miocene in Mexico (Baja California Sur). Evolution of a very large body size, decreased snout deflection, and loss of the tusks in *Dusisiren* suggests that these animals fed on kelp (order Laminariales) at or near the surface. *Dusisiren dewana*, known from 9-million-year-old rocks in Japan and California, is a likely intermediate between *D. jordani* and Steller's sea cow; it shows reductions in the number of teeth and finger bones.

In addition to *H. gigas*, other late-diverging *Hydrodama-*

Figure 5.3. Paleontologist Jorge Vélez-Juarbe excavating *Priscosiren* in Puerto Rico.

lis species include *H. cuestae* from late Miocene and Pliocene deposits in California and Mexico and *H. spissa* from the early Pliocene of Japan. *H. cuestae* remains have been found in California, Baja California, and Japan. This species was very large and might have reached 10 metric tons in weight. One specimen found in the San Diego region represents an individual over 9 m (30 ft) in length, making it the largest known sirenian. In addition to its large, rotund body, *H. cuestae* lacked teeth, had a whale-like tail, and had short forelimbs that lacked finger bones. Limb bones were pachyostotic.

Steller's sea cow, named for its discoverer, the German naturalist Georg W. Steller, was a gigantic animal. It measured at least 7.6 m (nearly 25 ft) in length and was estimated to weigh between 4 and 10 metric tons. The sea cow was unusual in lacking teeth and finger bones. Steller's sea cow lived in cold waters near islands in the Bering Sea—in contrast to the distribution of other sirenians in tropical or subtropical waters—and in prehistoric times ranged from Japan to Baja California. Steller described the sea cow's 3- to 4-inch-thick blubber as tasting something like almond oil, which might have played an unfortunate role in its demise. Steller's sea cow quickly became a major food resource for Russian hunters and early explorers in the North Pacific. One report states

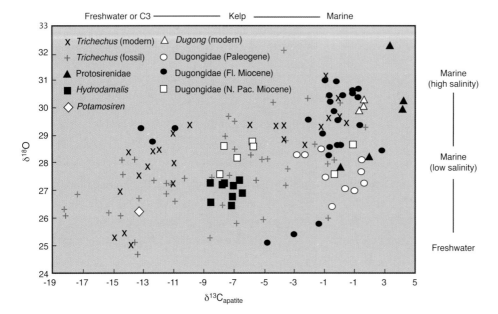

Figure 5.4. Carbon and oxygen isotope data for extinct and modern sirenians. From Clementz et al., 2009.

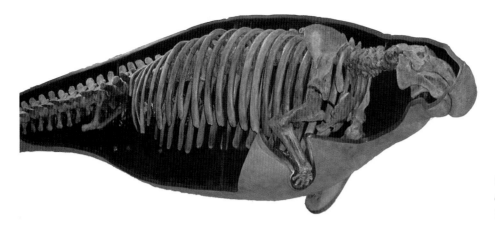

Figure 5.5. *Hydrodamalis cuestae* on exhibit at San Diego Natural History Museum.

that one sea cow could feed 33 men for a month. By 1768, only 27 years after its discovery, the sea cow was extinct. A recent study by Crerar and colleagues (2014) reports bones of *Hydrodamalis gigas* from St. Lawrence Island, also in the Bering Sea but further north and east, which date to 1,150-1,030 years ago. The samples were actually first discovered by dealers who were using the bones to craft knife handles. Nitrogen stable isotopes indicate a slight difference in feeding ecology than in the Bering Island population. Crerar et al. hypothesized that climate warming that would have changed kelp availability, coupled with aboriginal hunting by the newly arrived Inuit people, led to a second, earlier extirpation of a second historical population of *H. gigas* (see also chapter 7). Although raising an intriguing possibility for the pattern of extinction for Steller's sea cow, the results of this study have been questioned because they are based on uncatalogued specimens from the trade market, making it impossible to verify specimen identity and provenance. Scientists modeling the extinction of Steller's sea cow based on historical records (Bering and Copper Islands) and on life history data extrapolated from the closely related dugong suggest that the sea cows were greatly overexploited, hunted at seven times their sustainable limit. Population viability analysis indicates that the initial sea cow population was relatively small, estimated at 2,900 animals.

The *Metaxytherium* spp. + Hydrodamalinae clade is recognized as sister group to the Dugonginae, the subfamily to which the extant dugong belongs. Dugonginae includes, in addition to *Dugong*, the following extinct genera: *Bharatisiren, Callistosiren, Corystosiren, Crenatosiren, Dioplotherium, Domningia, Kutchisiren, Nanosiren, Rytiodus,* and *Xenosiren* (figure 5.2). Fossil remains of the dugongine clade have been found in late Oligocene rocks in the southeastern United States and the Caribbean and in other tropical waters of the Mediterranean, western Europe, Indian Ocean, South America, and the North Pacific.

The basal-most dugongine was *Crenatosiren olseni*, a small to medium-sized sea cow from the Caribbean region. The next-diverging *Nanosiren* lineage from the early Miocene to early Pliocene of the western Atlantic, Caribbean, Gulf of Mexico, and possibly eastern Pacific is distinguished by including the smallest sirenians, with adult body lengths of approximately 2 m (6.5 ft). The small tusks and strong snout deflection indicate that these animals foraged in shallower sea-grass beds (in nearshore waters) than the larger dugongids (e.g., *Metaxytherium*). Another basal tusked dugongid is *Callistosiren boriquensis* from the late Oligocene (26.51-24.73 million years ago) of Puerto Rico. The postcrania of *Callistosiren* differ from those of most other sirenians in being osteosclerotic but minimally pachyostotic, which is thought to be an adaptation for deeper diving in search of sea grasses, to depths greater than 10 m (32.8 ft). Rostral deflection (~43 degrees) and skull morphology suggest that this species may have fed on rhizomes of relatively large sea grasses in digging pits instead of feeding trails, as has been hypothesized for

Figure 5.6. Diorama at San Diego Natural History Museum showing *Hydroda-malis gigas* feeding in a kelp forest, with a nursing mother. Courtesy of W. Stout.

similar dugongids. In addition to *Crenatosiren* and *Callis-tosiren*, a third basal dugongid includes members of the *Nanosiren* lineage from the early Miocene through early Pliocene of North and South America, the Caribbean, and the Gulf of Mexico. *Nanosiren* species are among the smallest post-Eocene sirenians, with adult body lengths and weights of approximately 2 m (6.5 ft) and 150 kg (330 lb). In addition to small body size, members of this lineage had small conical tusks and a strong rostral deflection, indicating that they probably fed in nearshore marine waters on small sea grasses such as *Halodule* and *Halophila*.

The most elaborate development of tusks in the Sirenia is found in later-diverging dugongines such as *Bharatisiren*, *Rytiodus*, *Corystosiren*, *Domningia*, *Kutchisiren*, *Xenosiren*, and *Dioplotherium*. Enlarged, blade-like, self-sharpening tusks evolved in these species, probably for

digging up the rhizomes of larger sea grasses. Large tusks may have evolved in the modern dugong for a similar reason, but they seem to be used chiefly for social interactions. *Dioplotherium* is known from Brazil as well as California and Mexico, indicating dispersal through the Central American seaway. Another taxon providing evidence of the remarkable adaptive radiation of dugongines worldwide is *Bharatisiren*, related to *Dioplotherium* and found in India. One of the more interesting dugongines is *Corystosiren varguezi* from the Pliocene of Mexico and Florida. Its name comes from the Greek *korystos*, meaning "helmeted," in reference to its exceptionally thick skull roof consisting of 4.5 cm (more than 1.5 in) of solid, dense bone with no obvious functional explanation. Another taxon, *Xenosiren*, is hypothesized to be a direct descendant of *Dioplotherium*. It probably employed its large, self-sharpening tusks for digging for rhizomes. The discovery by Australian paleontologist Erich Fitzgerald and colleagues (2013) of a fossil dugongine from the early to middle Miocene (17.5–11.8 million years ago) in the Indian Ocean—the earliest Australasian record of sirenians—was to be expected, given the presence of living *Dugong* in that region today.

The living dugong, *Dugong dugon*, is distinguished by several derived characters: the absence of nasals, the consistent presence in juveniles of a deciduous first incisor, the frequent presence in adults of vestigial lower incisors, sexual dimorphism in size, eruption of permanent tusks (first incisors), functional loss of enamel crowns on the cheek teeth, and persistently open roots of the upper and lower second and third molars. Although the fossil record of the modern species is unknown, the fossil remains of a close relative were found in the late Pliocene of Florida, which supports a close relationship between *Dugong* and Caribbean fossil dugongines.

Sympatric Communities

Although the world now has only a single species of sirenian, this was not always the case. According to the fossil record of sirenians, it was more typical to find three or more species living together at one time. This co-occurrence of species, or sympatry, suggests that in regions where different sirenian species co-occurred, they had different feeding specializations in order to partition the feeding niches. In 2012, predoctoral researcher Jorge Vélez-Juarbe (currently at the Los Angeles County Museum of Natural History), Howard University paleontologist Daryl Domning, and Smithsonian paleontologist Nick Pyenson examined three dugongid localities from separate time periods: the late Oligocene in Florida, the early Miocene in India, and the early Pliocene in Mexico. All three localities showed fossil evidence that two or more sirenian species had once coexisted. In the Florida assemblage, tusk morphology was the dominant trait for separating feeding preferences. In contrast, in both the Indian and Mexican assemblages, multiple morphological traits (i.e., small body size and strong rostral deflection, the latter an adaptation for bottom feeding in shallower waters) separated co-occurring dugongids (figure 5.7). Several other sympatric sirenian communities have

been hypothesized, such as the coexistence of "halitheriine" dugongids *Priscosiren*, *Caribosiren turneri*, and *Caribosiren olseni* in Puerto Rico and South Carolina. Although these taxa differed little in body size they differed significantly in snout deflection and possibly in tusk shape, characters that suggest niche partitioning. Another sympatric community is suggested for *Metaxytherium albifontanum* (at least in part of its range) and two other dugongids, *Crenatosiren olseni* and *Dioplotherium manigaulti*, that differed in tusk size, which together with other cranial characters is hypothesized to serve as proxy for feeding preferences. *M. albifontanum*, with its small tusks, was less well suited to uprooting larger sea grasses than the other two species.

Evolution of Feeding

Stem sirenians (prorastomids) and the earliest dugongids are characterized by relatively horizontal rostra, suggesting that they probably fed on aquatic plants and, to a lesser degree, on sea grasses. Protosirenids, with a greater rostral deflection (35 degrees), probably fed on sea grasses, the primary food of modern dugongs. Oligocene faunas in several regions are typified by sympatric communities with as many as six co-occurring dugongid species. Miocene dugongids (e.g., *Metaxytherium* + Hydrodamalinae lineage) had downturned snouts and small to medium-sized tusks, indicating that they consumed seagrass leaves and small rhizomes. Members of the hydrodamaline lineage, Steller's sea cow in particular, with its loss of teeth and finger bones and its stout, claw-like forelimbs, are hypothesized to have fed on plants growing at the surface, using their claws to pull themselves forward as they grazed in shallow waters.

Trichechidae: Manatees

As recently as the nineteenth century, some scientists considered the manatee to be an unusual tropical form of walrus; in fact, the walrus was once placed in the genus *Trichechus* along with the manatees. The family Trichechidae includes the extant manatees, Trichechinae, and the extinct Miosireninae, a northern European clade composed of two genera: *Anomotherium* from the late Oligocene and *Miosiren* from the Miocene (figure 5.8). These

Figure 5.7. Sympatry in dugongid communities from Florida, India, and Mexico during the past 26 million years. From Vélez-Juarbe et al., 2012.

taxa were characterized by heavily reinforced palates that may have been adapted for crushing shellfish. The trichechid clade as a whole appears to have been derived from late Eocene or early Oligocene dugongids or from protosirenids.

The subfamily Trichechinae first appeared in the Miocene. *Potamosiren magdalenensis* is known from freshwater deposits in Colombia, a finding consistent with tooth isotope values for this taxon. Much of the manatees' history was spent in South America, from where they spread to North America and Africa in the Pliocene or Pleistocene. The unique type of tooth development in extant manatees, *Trichechus* spp., is termed horizontal replacement (although not the same as horizontal replacement observed in proboscideans), in which as worn teeth fall out at the

front of the jaw, new teeth are erupting in the rear. This mode of development did not evolve before the late Miocene taxon *Ribodon*. Based on morphological and some molecular data, the West African manatee, *Trichechus senegalensis*, and the West Indian manatee, *Trichechus manatus*, shared a more recent common ancestor than either shared with the Amazon manatee, *Trichechus inunguis* (figure 5.8). Two subspecies of the West Indian manatee can be distinguished on the basis of morphology and geography: the Antillean manatee, *Trichechus manatus manatus*, and the Florida manatee, *Trichechus manatus latirostris*. Extant manatees range from freshwater to marine habitats.

The majority of fossil remains of Pleistocene manatees from North America are referred to the extant species *Trichechus manatus*. During the late Pleistocene (late Rancho-

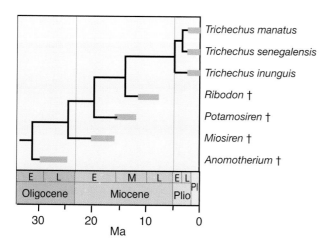

Figure 5.8. Phylogeny of Trichechidae. Modified from Domning, 2005.

labrean) a morphologically distinct subspecies, *Trichechus manatus bakerorum*, ranged from Florida to North Carolina. This fossil subspecies was found to be positioned close to the divergence of *T. manatus* and *T. senegalensis*. Daryl Domning (2005) hypothesized that warm intervals during the Quaternary allowed manatees from the Caribbean (currently represented by *T. m. manatus*) to disperse northward into the United States. Ecological barriers (cool winters on the northern Gulf of Mexico coast; deep water and strong currents in the Straits of Florida) impeded contact with Caribbean populations and permitted evolution of endemic North American forms (*T. m. bakerorum* and the living *T. m. manatus*). The African manatee, *T. senegalensis*, is hypothesized to have originated by dispersal from the New World to Africa.

Manatees are united as a monophyletic clade by features of the skull (e.g., the ear region). Other derived characters include reduction of the neural spines on vertebrae, a possible tendency toward enlargement and, at least in *Trichechus*, anteroposterior elongation of the thoracic vertebral centra. Mitochondrial DNA (mtDNA) sequence data support close divergence times for the three manatee species. Additionally, the genetic evidence suggests four lineages of *Trichechus* in the New World, one inhabiting the freshwater Amazon River, the other three inhabiting coastal systems and associated inland waterways. Studies of mtDNA have confirmed the genetic com-

plexity of various *T. manatus* populations. Analyses found *T. manatus* paraphyletic and closely related to *T. inunguis*. Further work on this is needed because genetic data support *T. inunguis* as a basal species, though the molecular clock and coalescent time calculations suggest that the lineage leading to *T. inunguis* evolved more recently.

Desmostylia

Desmostylians constitute the only extinct order of marine mammals. They have been closely allied with Sirenia and Proboscidea, together forming the Tethytheria. The fossil record indicates that divergence among the Tethytheria occurred near the Paleocene-Eocene boundary. If desmostylian origins lie within the Afrotheria (see chapter 1), from a stem proboscidean or anthracobunid (stem perissodactyl), there is a ghost lineage leading to the Desmostylia that extends from 57.9 to 49.0 million years ago. Anthracobunids, which have been suggested as the most likely candidate for sister taxon to the Desmostylia (e.g., Gheerbrant et al., 2005), were found in 40.4- to 55.8-million-year-old rocks in Asia. While this shortens the gap slightly, there is still 6-7 million years of missing record and nothing between the last record of anthracobunids in Pakistan and the record of desmostylians, which are from the North Pacific coast. There is also the possibility that desmostylians and anthracobunids are not Afrotherians but instead are allied with another group such as perissodactyls, which makes some biogeographic sense given the presence of early perissodactyls in both Asia and North America in the Eocene and Oligocene.

Desmostylians were confined to temperate latitudes in the North Pacific—Japan, Kamchatka (in the Russian Far East), and North America—during the late Oligocene and middle Miocene epochs, approximately 33-10 million years ago (figure 5.9). Known fossils of desmostylians include teeth, skulls, and skeletons of at least 11 genera and species (reviewed in Beatty, 2009), all of which were hippo-sized, amphibious quadrupedal browsers that probably fed on marine algae and sea grasses in subtropical to cool temperate waters. Desmostylians are thought to have been semiaquatic because they are found exclu-

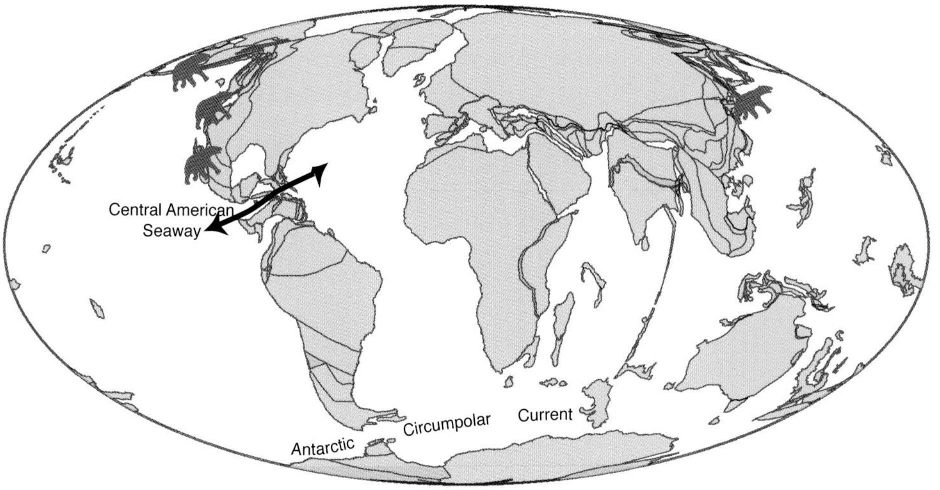

Figure 5.9. Major localities of desmostylians during the early Miocene to Pliocene. Modified from Fordyce, 2009. Base tectonic map from www.odsn.de/odsn/about.html.

sively in marine sediments, supporting an aquatic habitat, but at the same time their large fore and hind limbs are suggestive of terrestrial locomotion. Cranial features such as retracted nasals and raised orbits are additional features characteristic of semiaquatic habits.

The skull of desmostylians in general is massive, with a broad snout usually bearing procumbent canine teeth and incisor tusks and with a long diastema behind the canine teeth. The retracted nasals are a feature shared with proboscideans and sirenians. The molars display the characteristic closely bundled columnar cusps.

The first desmostylian fossil, from California, was described by the noted vertebrate paleontologist Othniel Charles Marsh in 1888. Marsh proposed the name *Desmostylus*, from the Greek, *desma*, meaning "bundle," and *stylos*, "pillar," referring to the bundled columnar shape of the cusps of the molar teeth in some taxa.

Desmostylian taxonomy has been hampered by scarcity, nonoverlapping elements, and polymorphism resulting from the different ontogenetic ages of specimens, necessitating revision of our understanding based on more complete material. The earliest-diverging desmostylians are represented by *Behemotops* (named by Linnaeus from the Biblical monster Behemoth), currently known by the species *B. proteus* and *B. katsuiei* from the middle or late Oligocene of North America and Japan. A new, partly articulated skeleton from Vancouver Island, British Columbia, is referred to *B.* cf. *B. proteus* (figure 5.10). Reevaluation of specimens originally designated as

Behemotops emlongi found they were not *Behemotops*, and the species was renamed *Seuku emlongi*. *Seuku* was named for a mythological hero of the native peoples of coastal Oregon (Alsea), from the same region as the holotype. Additional material for *Seuku* is needed to confirm its phylogenetic placement, although it has been suggested to fill the gap between *Behemotops* and the Paleoparadoxiidae. New information on the rostral anatomy of *Behemotops* suggests that rather than having a wide, flat-edged rostrum, as originally suggested, it may have had a narrow rostrum (figure 5.11). These basal taxa are relatively small and possess numerous ancestral characters.

Cornwallius, a later-diverging genus originally described from the late Oligocene of British Columbia, is the most geographically widespread desmostylian taxon—found in several eastern North Pacific, late Oligocene localities as far north as Alaska and as far south as Baja California Sur. *Cornwallius* is distinguished by its more strongly downturned upper canines, a postorbital process of the jugal, a dorsal midsagittal keel of the rostrum, paraglossal crests of the maxillary diastema, and a sagittal crest (figure 5.12.) Basally positioned relative to *Cornwallius* is the monotypic genus *Ashoroa* from marine strata of the Morawan Formation in Hokkaido, Japan. It is the smallest known desmostylian, with an estimated body length of 271 cm (8.9 ft).

A new genus of desmostylid, *Ounalashkastylus* from the Miocene of Alaska, has been described as more derived than *Cornwallius* but less derived than *Desmostylus*

Figure 5.10. Two male *Behemotops proteus* engaged in a dispute on the coast of Vancouver Island. Background based largely on photographs (by Beatty) of the modern coastline of Sombrio Beach, Vancouver Island, British Columbia, where specimens were collected, which is similar in flora and physiography to the present time. Illustrated by C. Buell.

and *Vanderhoofius*. These later-diverging desmostylids are thought to have suction-fed on sea grasses, on the basis of tooth morphology, a vaulted palate, and a medial eminence on the horizontal ramus that would have acted as a buttress to hold the jaws tight while the animal was feeding. *Vanderhoofius* was named by Roy Reinhart in 1959. Recent study of specimens with a more complete dental ontogenetic sequence for *Desmostylus* suggests that *Vanderhoofius* is a junior synonym of *Cornwallius*.

Paleoparadoxiids, the largest of the desmostylians, include three genera with four species: *Paleoparadoxia*, *Archaeoparadoxia*, and *Neoparadoxia*. The oldest record is from the latest Oligocene (ca. 24 million years ago) Skooner Gulch Formation, Point Arena, northern California. This specimen, a skeleton with skull, originally described as a species of *Paleoparadoxia*, is now referred to the genus *Archaeoparadoxia* as *A. weltoni*. As currently recognized, the genus *Paleoparadoxia* is known by two Miocene species. The older taxon, *P. tabatai*, is from the late early Miocene (ca. 18 million years ago) Aoki Formation in Honshu, Japan. A new specimen referred to *Paleoparadoxia* sp., reported from Hokkaido, Japan, from the early Miocene Sankebetsu Formation, is the geochronologically oldest known taxon. This taxon appears to be asso-

Figure 5.11. Reconstruction of the head of *Behemotops* cf. *B. proteus*. From Beatty and Cockburn, 2015. Illustrated by C. Buell.

ciated with warm temperate molluscan faunas, as is typical for *Paleoparadoxia* species, as well as with a cool-water temperate molluscan assemblage; this suggests a wider range of habitat tolerance than previously realized. A relatively large, later-diverging taxon, originally described as *P. repenningi* but reassigned to the genus *Neoparadoxia* (Barnes, 2013), is known by a single specimen (the "Stanford skeleton") of the middle Miocene (14 million years ago) Ladera Sandstone in northern California. In addition to *Archaeoparadoxia*, another large paleoparadoxiine, *Neo-*

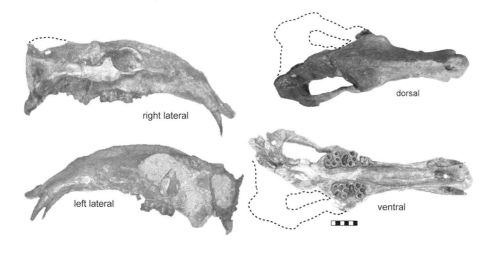

right lateral

dorsal

left lateral

ventral

Figure 5.12. Skull of *Cornwallius sookensis* in dorsal, ventral, and lateral views. Scale bar = 6 cm. From Beatty, 2009.

Figure 5.13. Life restoration of *Neoparadoxia.* From Barnes, 2013.

paradoxia cecilialina, has been described based on a skeleton of late Miocene age from the Monterey Formation of southern California. Reconstruction of the holotype skeleton of *N. cecilialina* indicates that this individual, a juvenile, was approximately 2.2 m (7.2 ft) long and could have attained an adult body length of 2.73 m (8.9 ft), making it the largest known paleoparadoxiid. The body of this species is reconstructed as stout, with short, heavy limbs and large pelvic bones angled posteroventrally, anteriorly protruding incisors, large nostrils, and elevated eyes (figure 5.13). It probably used alternate forelimb paddling, similar to that of polar bears, with palms facing posteromedially when abducted beneath the trunk. *N. cecilialina* has been envisioned as feeding on the seafloor, with its hind limbs stabilizing its body. During paleoparadoxiid evolution, the oral cavity became larger and the snout

and mandible were increasingly ventrally turned, resulting in the mandibular symphysis, lower canines, and lower incisors being nearly horizontally oriented. Such ventral turning also occurs among dugongid sirenians, in which it has been interpreted as an adaptation for bottom feeding on aquatic vegetation. In paleoparadoxiids, ventral turning might likewise have been an adaptation for bottom feeding.

Desmostylus, the most specialized and best-represented genus of the order, is found widely in Miocene coastal deposits of the North Pacific. It exhibits the distinctive molars that gave the genus its name. A study of isotopic concentrations in tooth enamel in *Desmostylus* by Mark Clementz and colleagues (2003) found that these sea cows had a diet with high ^{13}C values, suggesting a diet of submerged aquatic vegetation. Coupled with the [18]O val-

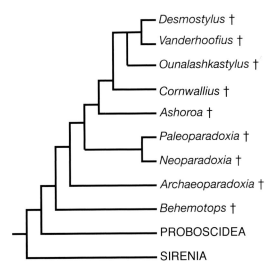

Figure 5.14. Phylogeny of desmostylians. Modified from Beatty, 2009.

ues, this indicated that *Desmostylus* spent as much time in water as a pinniped. Strontium isotope values were similar to those for terrestrial rather than marine mammals, suggesting that this animal was an aquatic herbivore that foraged in estuarine or freshwater systems. Dental material exhibiting an ontogenetic sequence indicates that development of the mandible and the tooth eruption sequence in *Desmostylus* are similar to the ontogeny (delayed eruption) characteristic of Afrotheria, further supporting a hypothesized shared ancestry. Like paleoparadoxiids, *Desmostylus* has been shown to be a forelimb-dominated swimmer.

A phylogenetic analysis of desmostylians strongly supports a clade comprising *Desmostylus*, *Vanderhoofius*, *Cornwallius*, *Ashoroa*, *Neoparadoxia*, and *Paleoparadoxia*, with *Archaeoparadoxia* and *Behemotops* (and likely sister *Seuku*) as consecutive sister taxa (figure 5.14). Synapomorphies that unite desmostylians include transversely aligned lower incisors, an enlarged passage through the squamosal from the external auditory meatus to the roof of the skull, fused roots of the lower first premolar, and an elongated paroccipital process.

Reconstructions of the skeleton and resulting inferences regarding the locomotion of desmostylians have been both controversial and amusing. Various renditions of these animals have included resemblances to sea lions,

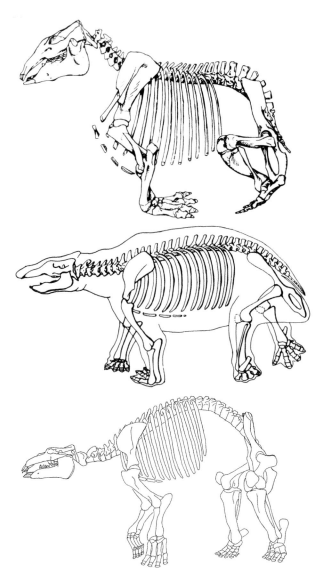

Figure 5.15. Desmostylian reconstructions. From Inuzuka, 1984; Domning, 2002.

frogs, and crocodiles. Studies by Domning (2002) suggest that desmostylians had an upright posture similar to that seen in some ground sloths and chalicotheres. The body was probably well off the ground, with the limbs under the body (figure 5.15). The hind limbs were heavy as judged by the large femur and hip joint, with vertical ilium, abducted knees, and digitigrade foot posture. It has been suggested that desmostylians may have supported their weight on their hind quarters as they pulled vegetation from overhanging branches. They are thought to have been slow-moving, scrambling on uneven and slippery surfaces while moving between land and sea. Loco-motion in the water was by forelimb propulsion resembling that of polar bears, with which they share similar limb proportions. Dental morphology is varied, with later-diverging species showing adaptations for an abrasive diet, probably one that contained grit mixed with plant material scooped from the sea bottom or shore. A stable isotope study of tooth enamel from *Desmostylus* suggests that these mammals spent time in estuarine or freshwater environments rather than exclusively marine ecosystems and probably foraged on sea grasses as well as a wide range of other aquatic vegetation.

A study of bone histology suggests that *Behemotops*, *Paleoparadoxia*, and *Ashoroa* exhibited an increase in bone volume and compaction (i.e., osteosclerosis and pachyosteosclerosis) consistent with being shallow-water swimmers that either hovered slowly at a preferred depth or walked on the bottom. Conversely, *Desmostylus*, which differs from these desmostylians in possessing a spongy inner bone organization (osteoporotic-like pattern), is thought to have been a more active swimmer that may have fed more at the surface (Hayashi et al., 2013).

Oligocene desmostylians *Cornwallius sookensis*, *Behemotops proteus*, and *Seuku emlongi* apparently coexisted in the Pacific Northwest. This provides another example of a marine mammal sympatric community, paralleling that of sea cow assemblages.

Aquatic Sloths and Recent Occupants of the Sea, Sea Otters and Polar Bears

Swimming Sloths

Today, only two groups of sloths exist, and both live in the trees. During the Miocene, a greater diversity of sloths lived not just in trees but also on the ground and in water. One of the more extraordinarily marine-adapted mammals is a lineage of megatheriid "ground sloths" (Amson et al., 2016). As many as five species of the extinct marine sloth *Thalassocnus* have been described from the coasts of Peru by French vertebrate paleontologist Christian de Muizon and colleagues (e.g., Muizon and McDonald, 1995; Muizon et al., 2003, 2004b) and, more recently, from the late Miocene to late Pliocene (7.2–2.5 million years ago) of Chile, based on abundant associated complete and partial skeletons (figure 6.1). In Peru each species was found in a different level of rock unit in the Pisco Formation, and their progressive and differing aquatic adaptations suggest that these species may represent a single anagenetic lineage. The Peru sample is from one of the few sites where a hypothesized ancestor-descendant lineage is present in one location and in a single stratigraphic sequence.

The Pisco Formation corresponds to a major marine transgression along the southern coast of Peru during the middle Miocene to late Pliocene. Each species of *Thalassocnus* is found in a distinct stratigraphic horizon. The oldest species, *T. antiquus*, from the Aguado de Lomas Horizon, is approximately 7.8 million years old; *T. natans*, from the Montemar Horizon, 6 million years old; *T. littoralis*, from the Sud-Sacaco Horizon, 3 million years old; *T. carolomartini*, from the Sacaco Horizon, 3–4 million years old; and *T. yaucensis*, from a late Pliocene Horizon, 1.5–3 million years old. A large sample of *Thalassocnus* specimens preserves adult as well as juvenile individuals that encompass a considerable size range, from small species such as *T. littoralis* with a femoral length of 265 mm (10 in) to the large *T. yaucensis* with a femur approximately two-thirds larger.

The most distinctive feature of the skull, one that distinguishes *Thalassocnus* from other nothrotherine sloths, is elongation of the anterior part of the skull and modifica-

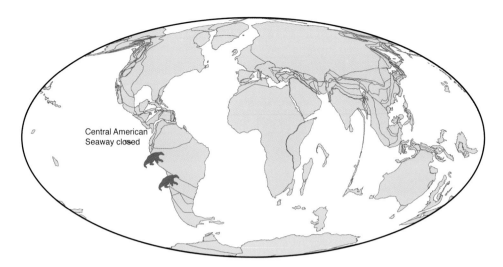

Figure 6.1. Localities of aquatic sloths during the late Miocene to Pliocene. Modified from Fordyce, 2009. Base tectonic map from www.odsn.de/odsn/about.html.

tion of the premaxilla and its occluding mandibular symphysis. In addition, the snout and lower jaw are laterally expanded so that they become spatulate. The anterior skull undergoes modification in the successive species: in earlier species (*T. natans*) the premaxilla is strongly deflected, then in a later-diverging species (*T. carolomartini*) the premaxilla shows virtually no ventral deflection, presenting a straight skull shape. The change in angle of the premaxilla relative to the skull and consequent straightening of the skull, which affects feeding, is related to the more posterior projection of the condyles in *T. littoralis* than in *T. natans*; these modifications have been associated with an increasing aquatic specialization in the lineage also seen in later-diverging species. Modifications in the two youngest species include an anterior widening of the premaxilla and a splayed apex of the mandibular spout (anterior-most end of the mandibular symphysis). Another feature related to feeding (and breathing) while the head is partially submerged is the anterodorsally facing narial opening (as also seen in phocids) in the later-diverging species *T. littoralis*, *T. carolomartini*, and *T. yaucensis*.

The crushing, worn molars and downturned rostrum with an expanded tip suggest the presence of well-developed lips for grazing on sea grasses. Based on reconstruction of the feeding habits of *Thalassocnus*, Muizon et al. (2004a) concluded that the early-diverging species of the genus (*T. antiquus*, *T. natans*, *T. littoralis*) were par-

tial grazers (intermediate or mixed feeders), as indicated by abundant dental striae on the teeth created by the ingestion of sand. Later-diverging species are more likely to have been specialized grazers, with a distinctive horizontal component in the mandibular movement (figure 6.2). The shallow dental basins of the teeth of *T. yaucensis* also indicate more efficient anterior and posterior movements with the mouth closed, which is compatible with efficient grazing. In addition to *Thalassocnus*, desmostylians and sirenians have convergently evolved adaptations for feeding on marine vegetation. Desmostylians exhibit efficient tools for grazing, such as premaxillae with expanded tips and wide mandibular symphyses, whereas sirenians lack expanded premaxillae but have wide upper lips and horny pads on the maxillae and mandibles that are used to grasp sea grass during feeding.

A thorough description and functional interpretation of the postcranial skeleton of *Thalassocnus* was recently reported by paleontologist Eli Amson and colleagues (Amson et al., 2014, 2015a, b). Forelimb features such as the shortness of the humerus and radius and the specialized robust digits with large claws distinguish *Thalassocnus* from other sloths. Moreover, the later-diverging species of the genus (e.g., *T. carolomartini*, *T. yaucensis*) are characterized by development of the pronator ridge of the radius, stoutness of the ulna, widening of the proximal carpal row, and shortening of the metacarpals (figure 6.3). It is hypothesized that, in addition to quadrupedal

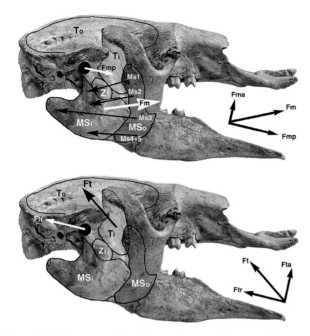

Figure 6.2. Reconstruction of muscular attachments of masseter (*top*) and temporalis (*bottom*) in skull of *T. yaucensis. Abbreviations:* Fm, general force produced by the action of the masseter pars profundus; Fma, adduction force of the masseter pars superficialis; Fmp, protraction force produced by the masseter pars profundus; Ft, general force produced by the action of the temporalis; Fta, adduction force of the temporalis; Ftr, retraction force of the temporalis; MSi, insertion of the masseter pars superficialis; MSo, origin of the masseter pars superficialis; Ti, insertion of the temporalis; To, origin of the temporalis muscle; Zo, origin of the zygomaticomandibularis. From Muizon et al., 2004a.

paddling, the forelimb of *Thalassocnus* was involved in bottom walking. The forelimb of *Thalassocnus* was also probably used in activity related to obtaining food such as uprooting sea-grass rhizomes.

The horizontal orientation and weak development of the iliac wing of the pelvis in *T. carolomartini* compared with earlier species of the genus may reflect its aquatic habits and need for support of the body in water, rather than the need to counter the effects of gravity on land. In the hind limb of *Thalassocnus*, the tibia is longer relative to the femur, features usually found in aquatic and semi-aquatic taxa. Comparison of limb ratios among *Thalassocnus* species indicates that the hind limb was smaller in later-diverging species. Morphology of the vertebral column of *Thalassocnus* does not suggest that it was used

in propulsion, and the increased stability of the posterior portion of the column (e.g., robust zygapophyseal articulations) is instead indicative of underwater digging. The tail of *Thalassocnus* is relatively long and wide for most of its length due to the greater number of caudal vertebrae (more than 20)—more than one-third longer than in other sloths. The tail was muscular as evidenced by well-developed bifurcated transverse processes and was probably used for diving and providing underwater stability rather than for propulsion.

The five species of *Thalassocnus* document the shift from one posture to another. The stance with weight borne on the lateral side of the foot was seen in the earliest species of *Thalassocnus*, as in other megatheriid ground sloths (figure 6.4). This stance was apparently forsaken by later species of the genus in favor of a secondary plantigrady, with the weight borne on the sole of the foot. Quadrupedal swimming was probably a dominant

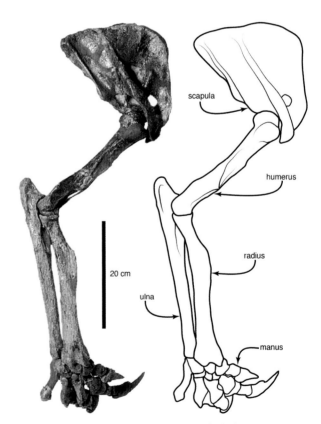

Figure 6.3. Articulated right forelimb of *Thalassocnus natans*. From Amson et al., 2015b.

Figure 6.4. Life restoration and lateral views of the skeleton of *Thalassocnus natans*. From Amson et al., 2015a. Life restoration illustrated by C. Buell.

swimming mode of *Thalassocnus*, and a plantigrade hind limb and enlarged claws may have been more efficient for paddling and for bottom walking.

In the earliest species of *Thalassocnus*, osteosclerosis and pachyostosis are absent, but the ribs and limb bones of later-diverging species such as *T. carolomartini* show gradual acquisition of osteosclerosis. This increasing bone thickness suggests an adaptation for buoyancy with more time spent submerged.

Although rare, dugongine sirenians were previously reported in the Pisco Formation and suggested as potential competitors with *Thalassocnus*. However, histological restudy of the "sirenian" rib demonstrated that it is clearly distinguishable from those of sirenians and matches that of *Thalassocnus*, which means that aquatic sloths did not stratigraphically overlap with sirenians in the Pisco Formation. Of significance is that the sea cow niche was occupied by *Thalassocnus* after sirenians became extinct in the southwest Atlantic. The similarity of these marine sloths to extinct desmostylian pinnipeds in the North Pacific in terms of diet raises the possibility that they may be the ecological homologues of desmostylians in the South Pacific. The extinction of marine sloths has been related to a decrease in sea temperature, which may have altered food sources and/or availability. The supposedly low basal metabolic rate of *Thalassocnus* (as seen in extant xenarthrans) is also possibly implicated in its extinction, as colder waters would have increased the metabolic cost of endothermy.

Marine Otters

Marine otters include the sea otter, *Enhydra lutris*, and a marine species of South American otter, *Lontra felina*. Both are members of the carnivore family Mustelidae, which also includes 70 species of river otters, skunks, and weasels, among others. Although sea otters are the smallest marine mammals, they are the largest mustelids (1.4 m [4.6 ft] in length). The generic name of the sea otter is from the Greek *enhydris* and the specific epithet is from the Latin *lutra*, both meaning "otter." Three subspecies of sea otter are recognized based on differences in cranial morphology as well as geographic distribution. The western

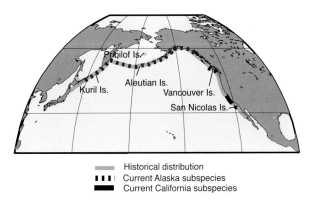

Figure 6.5. Sea otter distribution map. From Estes and Bodkin, 2002.

sea otter, *Enhydra lutris lutris*, inhabits the Kuril Islands, the east coast of the Kamchatka Peninsula, and the Commander Islands. The eastern sea otter, *Enhydra lutris kenyoni*, ranges from the Aleutian Islands to Oregon. The southern sea otter, *Enhydra lutris nereis*, had a historical range from northern California to approximately Punta Abrejos, Baja California, Mexico; today it occupies fragmented ranges in California, with populations centered in Monterey Bay and on San Nicolas island (figure 6.5).

Extant sea otters are primarily durophagous predators feeding mostly on various hard benthic invertebrates, notably sea urchins, clams, and abalones. Certain Alaskan stocks and western sea otters, however, incorporate a greater proportion of fish in their diets. Considering the difference in geographic location and diet among the three subspecies, cranial variation among the subspecies was predicted. Morphometric study by functional anatomist Timm-Davis and colleagues (2015) revealed both a smaller mechanical advantage for a major jaw-closing muscle (masseter) and smaller bite forces in western sea otters than in the other subspecies, corresponding with this dietary difference. Sea otters typically pry prey off their substrates using forepaws and rock tools, then crush the shells with their teeth or, using tools, against their chests before consuming the prey. Although sea otters are able to walk on land they spend most of their time in water, resting in kelp beds. Extant sea otters are primarily hind limb-dominated swimmers, but this was not true for fossil species.

The modern sea otter, *Enhydra*, arose in the North Pacific at the beginning of the Pleistocene, about 3–1 million years ago, and has not dispersed since that time. Records of *Enhydra* exist from the early Pleistocene of Oregon and California and the Plio-Pleistocene of Alaska. One extinct species, *Enhydra macrodonta*, has been described from the middle Pleistocene of California. The proportions of the mandible suggest that the skull of this species may have been slightly larger than that of the extant sea otter. Another species, *Enhydra reevei*, based on a few teeth from the late Pliocene of the United Kingdom, is more *Enhydra*-like than any *Enhydritherium* or *Enhydriodon* described and, according to Willemsen (1992), is the oldest species of *Enhydra*.

The closest living relatives of *Enhydra* are other lutrine otters. The lutrine otters are classified into two groups based on the morphology of the carnassial teeth (upper fourth premolar and lower first molar). One group, the "fish-eating" otters, includes the extinct otter genus *Satherium* and the extant otter genera *Lutra* (Eurasian otter), *Lontra* (American river otter), *Hydrictis* (spotted-neck otter), *Aonyx* (short-clawed otter), *Pteronura* (giant otter), *Amblonyx* (oriental small-clawed otter), and *Lutrogale* (smooth-coated otter). These otters are characterized by more blade-like carnassials than the other group of otters, the bunodont otters.

Bunodont otters have non-blade-like carnassials with thick enamel and rounded cusps. The bunodont otters include the extant genus *Enhydra* (sea otters) as well as the extinct giant otter genera *Enhydritherium* and *Enhydriodon*. In a phylogenetic analysis that included both the living sea otter and related extinct taxa, Berta and Morgan (1985) proposed two lineages of sea otters: one early-diverging lineage that led to the extinct genus *Enhydriodon* and a later-diverging lineage that led to the extinct giant otter, *Enhydritherium*, and the extant sea otter, *Enhydra*. *Enhydriodon* is known only from Africa and Eurasia, from sediments of the late Miocene to late Pliocene (5–0.7 million years ago). We do not know whether *Enhydriodon* lived in marine or freshwater habitats or both. These animals were as large as or larger than modern sea otters and had a similarly well-developed molariform dentition. It has been suggested that the diet consisted mostly of hard food such as molluscs and catfish. The gigantic "bear otter," *Enhydriodon dikikae*, from the Pliocene (4 million years ago) of Dikika in the lower Awash valley of Ethiopia is distinguished by, in addition to its large size, its short snout, large robust teeth, and long powerful canines (figure 6.6). This species is estimated at 77–126 kg (169–277 lb) with a body length of more than 2 m (7 ft). Postcranial remains suggest that *E. dikikae* was mostly terrestrial. For example, the femur is slender and the medial epicondyle of the humerus is weak in comparison with that of truly aquatic otters (*Pteronura*, *Enhydra*).

Enhydritherium is known from the late Miocene of Europe and the late Miocene to middle Pliocene (13.65–3.6 million years ago) of North America. Two species of *Enhydritherium* are described: *E. lluecai* (*Paludolutra* fide Pickford, 2007) from Spain and *E. terraenovae* from Florida and California. *Enhydritherium* is united with *Enhydra* based on dental synapomorphies. *Enhydritherium* may be distinguished from both *Enhydriodon* and *Enhydra* by the

Figure 6.6. Anterior skull and mandible (*left and middle*) and palate (*right*) associated with *Enhydriodon dikikae*. Scale bar = 10 cm. From Geraads et al., 2011.

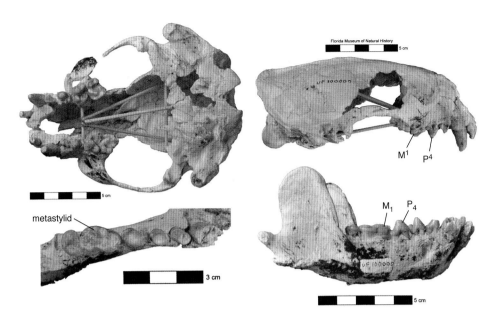

Figure 6.7. Skull (*top*) and lower jaw (*bottom*) of *Enhydritherium terraenovae* showing upper and lower dentition. *Abbreviations:* M, molar; P, premolar. Copyright © Florida Museum of Natural History, University of Florida.

anterior and medial position of the protocone in the upper fourth premolar and by absence of the protostylid and presence of the metastylid on the lower first molar (figure 6.7). The purported close relationship between Old World otters (e.g., *Enhydriodon*) and North American otters (e.g., *Enhydra* and *Enhydritherium*) may be the result of convergence, but this hypothesis requires reexamination in the light of newly reported material such as *Enhydriodon dikikae*.

An incomplete articulated skeleton of the North American sea otter, *E. terraenovae*, was described from the Moss Acre Racetrack site in northern Florida. The depositional environment of this site, which is located a considerable distance from the coast, indicates that *E. terraenovae* was probably more terrestrial than previously thought and frequented large inland rivers and lakes in addition to coastal marine environments. *Enhydritherium* differed from *Enhydra* in its forelimb-dominated swimming and skeletal specializations (i.e., more equally proportioned fore and hind limbs) suggesting that its locomotion on land was more proficient. *Enhydritherium* was similar in size to *Enhydra*, with an estimated body mass of approximately 22 kg (48 lb). The unspecialized distal hind-limb elements and heavily developed humeral muscles of *Enhydritherium* strongly suggest that, unlike *Enhydra*, this animal was primarily a forelimb swimmer. Given

its almost certainly more effective terrestrial locomotion than *Enhydra*, *Enhydritherium* is likely to have spent more time on land than the living species. The thickened cusps of the upper fourth premolars of *E. terraenovae* and their tendency to be heavily worn (figure 6.7) suggest that these otters, like *Enhydra*, consumed extremely hard food items such as molluscs. Other food items such as fish were also eaten, however, as suggested by the concentration of fish fossils (most likely stomach contents) around the Florida locality where the partial skeleton was found. The combined evidence of lack of specialization for hind-limb swimming, hand proportions more similar to those of modern fish-eating otters, and osteological evidence for powerful neck muscles suggests that *E. terraenovae* was well-adapted for catching fish and other moving prey with its jaws, rather than using its hands in prey manipulation as does *Enhydra* (Lambert, 1997).

Sea Mink

The extinct sea mink, *Neovison macrodon*, is best known from Native shell middens dating as far back as 5,100 years in marine and estuarine habitats along the coastal islands of the Gulf of Maine. It was originally included in *Mustela*, but cytogenetic and biochemical data support its inclusion, along with the American mink, *N. vison*, in the genus *Neovison* rather than *Mustela*. The fossil record of

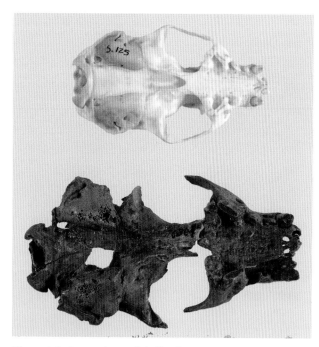

Figure 6.8. Comparison of skulls of *Neovison macrodon* (*bottom*) and *Mustela vison* (*top*). Courtesy of A. Spiess.

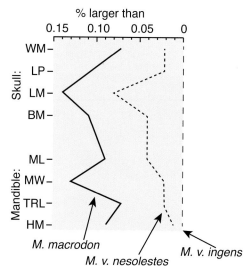

Figure 6.9. A log-log ratio diagram for cranial measurements and mandible measurements (of various subspecies of *Neovison* (*M.*) *vison* and *Neovison* (*M.*) *macrodon*. *Neovision* (*M.*) *v. ingens* is used as the comparative point (0). *Abbreviations:* BM, width at M¹; HM, mandible height; LM, length at M¹; LP, palate length; ML, mandible length; MW, mandible width; TRL, length of tooth row; WM, palate width at M¹. Modified from Mead et al., 2000.

this species is not well known, although the earliest occurrences date from the early Pleistocene (Irvingtonian), and the species is uncommonly recorded from the late Pleistocene (Rancholabrean). Of the approximately 25 Pleistocene occurrences of the sea mink, all are within the present-day range of the extant American mink, and there seems to be little morphological difference between Irvingtonian and Rancholabrean specimens. The marine species can be distinguished from the American mink by the larger overall size and robustness of the former, especially the teeth (figures 6.8, 6.9). The largest known specimen approximated 83 cm (nearly 3 ft) in length. Like that of the American mink, the diet of the sea mink consisted of seabirds. Locomotion in the sea mink is by quadrupedal paddling, differing from the hind limb–dominated propulsion of sea otters. The species has the distinction of being the only mustelid hunted to extinction, a victim of the fur trade in the late 1800s.

Polar Bears

The polar bear, *Ursus maritimus*, also known as the ice bear or sea bear, is the only species of bear that spends a significant portion of its life in salt water. The polar bear is the largest of six extant bear species and the most recently evolved marine mammal species, appearing perhaps less than 500,000 years ago in the Arctic. The generic name for the polar bear, *Ursus*, is the Latin for "bear," and its specific epithet, *maritimus*, refers to the maritime habitat of this species. The suggestion that the polar bear might represent a separate genus, *Thalarctos*, because of its adaptation to aquatic conditions and its physical appearance, is not supported.

Earlier molecular data suggested that polar bears diverged from brown bears between 1.5 and 1 million years ago. More recently, DNA analyses of remains of the oldest known polar bear fossil found north of the Arctic Circle provide an "evolutionary snapshot" revealing when and where polar bears evolved. These new data confirm that polar bears originated in the Arctic but provide evidence that they split more recently from brown bears, probably less than 500,000 years ago. Stable isotope analyses of the fossil teeth indicate that polar bears clearly had a different feeding ecology than brown bears

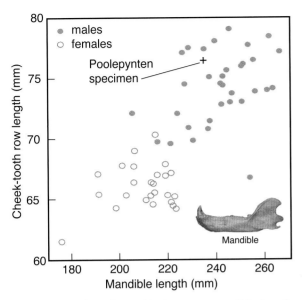

Figure 6.10. The relationship between mandible length and cheek-tooth row length in polar bears. Modified from Ingólfsson and Wiig, 2008.

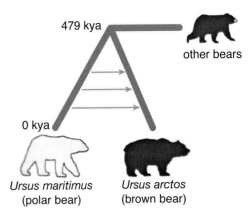

Figure 6.11. Polar bear and brown bear divergence. Based on Liu et al., 2014.

and that early in their evolutionary history, polar bears adapted rapidly to feeding on seafood such as fish and marine mammals, attaining their unique position as top marine predators in the Arctic.

The fossil record of polar bears is poor and limited to the Pleistocene. Polar bear fossils are known from southern England, northern Spain, and the North Sea. The oldest remains of a polar bear, a lower jaw from Svalbard, Norway, is dated at 130,000-110,000 years ago (figure 6.10). Diagnostic features of this polar bear and the size of the jaw suggest that it is from an adult male, similar in size to the living species. The cold, dry conditions of the fossil locality kept the DNA well preserved, and this fossil jaw provided the oldest mammalian mitochondrial genome yet sequenced—about twice the age of the oldest mammoth genome, which dates to around 65,000 years ago.

Despite extensive DNA analysis and a close alliance between polar bears and brown bears, *Ursus arctos*, the exact nature of their relationship remains unclear. Using population genomic modeling to analyze nuclear genomes of polar and brown bear populations, Liu et al. (2014) found that the species diverged only 479,000-

343,000 years ago, although prior divergence estimates ranged from 5-4 million to about 600,000 years ago (figure 6.11). Strong evidence exists for continuous gene flow from polar bears into brown bears after the species diverged, which is in agreement with the findings of Cahill et al. (2013). When these genomic data are coupled with stable isotope analysis of a canine tooth in the Pleistocene jawbone collected from Svalbard, the information indicates that polar bears were adapted to a marine diet and life in the high Arctic by at least 110,000 years ago (Lindqvist et al., 2010). This implies a very rapid evolution and suggests that the distinct adaptations of polar bears may have evolved over fewer than 20,500 generations.

Another interesting aspect to the genetic history of brown bears and polar bears comes from the study of a group of brown bears that live on the Admiralty, Baranof, and Chichagof (ABC) islands of southeastern Alaska. These bears have mtDNA that more closely matches that of polar bears than other brown bears. The most likely evolutionary history of the ABC brown bears suggests that they descended from polar bears that gradually evolved into brown bears through hybridization with male brown bears dispersing from the Alaskan mainland (Cahill et al., 2013). This hypothesis is consistent with present-day climate events. Polar bears in some areas are spending more time on land as a result of climate warming and reduced access to ice.

7 Diversity Changes through Time

The Influence of Climate Change and Humans

Global Patterns
Past Diversity

Global patterns and processes of species diversity are important factors that determine extinction risk and need to be considered in conservation planning. Marine mammal diversity, at the present and in the past, is regulated by either biological processes such as competition and predation (the Red Queen hypothesis) or environmental factors such as climate and food supply (the Court Jester hypothesis), or by a mixture of both. As reviewed by vertebrate paleontologist Michael Benton in 2009, it is clear that both processes drive evolution; the debate is about which is more important and when, and on what scale (geographic and temporal).

The Red Queen hypothesis is named for the Red Queen in *Through the Looking Glass*, who tells Alice that "it takes all the running you can do, to keep in the same place." This proposition is based on the idea that no animal (or other living organism) is alone, and if you don't evolve, the animals around you will, and you'll be doomed. The alternative hypothesis is that evolution is caused by environmental changes—the Court Jester hypothesis. In the context of marine mammal evolution, for example, the adaptive radiation of crown cetaceans has been explained in terms of the extinction of stem whales and the evolution of echolocation and baleen—key innovations that promoted new ecological opportunities and evolution of crown cetaceans, as discussed by University of Chicago evolutionary biologist Graham Slater and colleagues (2010). But physical factors may have been equally important in terms of organizing these radiations. For example, diversification of cetaceans at the end of the Oligocene is hypothesized to have resulted from physical restructuring of the oceans. This included initiation of the Antarctic Circumpolar Current, upwelling, and a subsequent increase in primary production (e.g., Steeman et al., 2009). Comparative phylogenetic approaches enable tests of the hypotheses

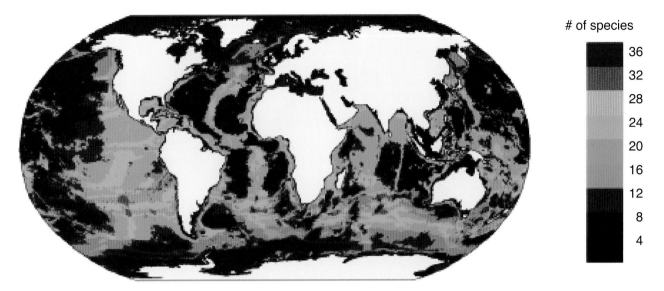

# of species

36
32
28
24
20
16
12
8
4

Figure 7.1. Predicted patterns of present-day marine mammal species richness. From Kaschner et al., 2011.

about macroevolutionary patterns and offer new insights that have only just begun to be explored.

Current and Future Diversity

Current patterns of marine mammal biodiversity that were analyzed based on an environmental suitability model (bathymetry, sea surface temperature) were found to compare well with observed patterns of species richness. The largest present-day concentrations of marine mammals are found in temperate waters of both hemispheres (Kaschner et al., 2011) (figure 7.1). Comparisons among different marine mammal lineages reveal, first, that mysticetes differ from odontocetes in being concentrated in mid-latitudes, and second, that pinnipeds are more highly concentrated in subpolar and polar waters.

Using this framework, scientists have generated predictions of future patterns of diversity. The predicted future (2040–2049) effects of environmental change on marine mammals in Kaschner et al.'s analysis were moderate. They revealed a decrease in some species (e.g., native species in the Barents Sea, northern Indian Ocean, waters surrounding Japan, waters around the Galápagos Islands) and an increase in other species, mostly through the invasion of new species in polar waters (e.g., northern

Greenland Sea, central Bering Sea, High Arctic waters, portions of the Weddell Sea) (figure 7.2). In particular, pinniped biodiversity in tropical and temperate waters was predicted to decrease substantially.

Climate Change and Its Effect on Distributions

Global warming is a reality and, in large part, is human-caused. The increase of carbon dioxide and other greenhouse gases in the atmosphere is thought to be the main cause of global climate warming. The impact of global warming on marine mammals has already been detected. Marine mammals have been described as sentinels of ecosystem change by National Oceanic and Atmospheric Administration (NOAA) scientist Sue Moore (2008), since they reflect ecological variation on large spatial and temporal scales. Among the changes observed are distributional shifts, habitat loss, and changes in diets and food webs in marine ecosystems.

Climate-Related Distribution Shifts: Whales and Fur Seals

Biogeographic shifts in response to climate change (e.g., those associated with glacial-interglacial cycles) have been shown for various marine mammal lineages. One

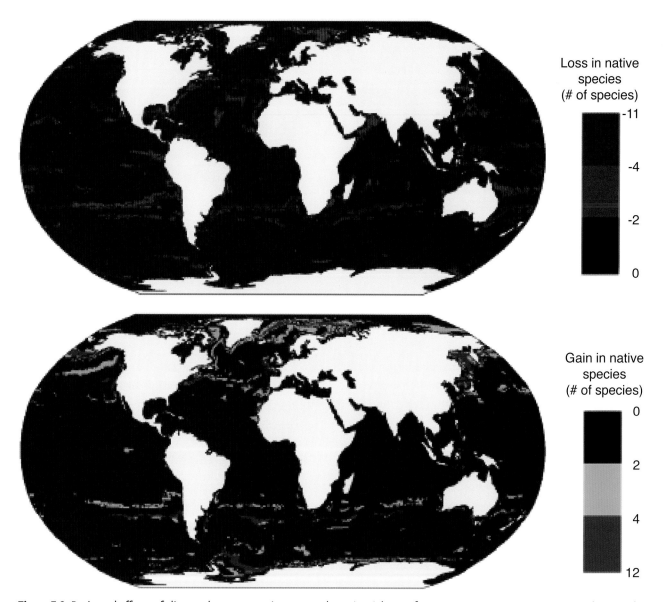

Loss in native
species
(# of species)

-11

-4

-2

0

Gain in native
species
(# of species)

0

2

4

12

Figure 7.2. Projected effects of climate change on marine mammal species richness, from 1990–99 to 2040–49. From Kaschner et al., 2011.

method of studying climate-related distribution shifts is through the analysis of ancient DNA (aDNA). Studies of aDNA have revealed the population dynamics of species responding to habitat as it expanded and contracted through glacial cycles. For example, in a study using aDNA analysis and habitat suitability modeling, researchers found that the bowhead whale, *Balaena mysticetus*, currently inhabiting Arctic waters, had shifted its range to more southerly latitudes during the Pleistocene-Holocene transition (Foote et al., 2013). Furthermore, at

the beginning of the Holocene as ice receded and warmer water predominated, the bowhead was replaced in those latitudes by its sister species, the North Atlantic right whale, *Eubalaena glacialis*, a temperate-water occupant (figure 7.3).

Another study of climate-related distribution shifts found that from 1984 to 2010, fin and humpback whales, *Balaenoptera physalus* and *Megaptera novaeangliae*, had shifted the timing of their arrival on a North Atlantic feeding ground (the Gulf of St. Lawrence). The analysis

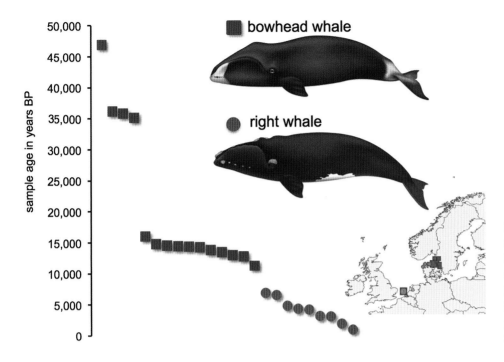

Figure 7.3. Changes in species occurrence—bowheads replaced by right whales (see text)—during the Pleistocene-Holocene transition (in years before the present). From Foote et al., 2013.

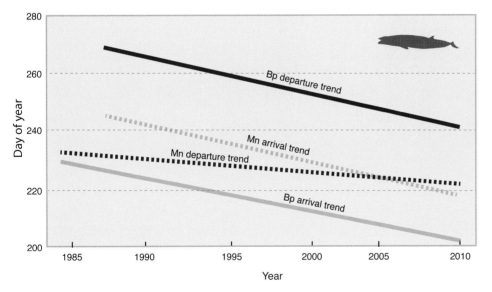

Figure 7.4. Mean annual arrival date of fin whales (Bp) and humpback whales (Mn) the first week that the Gulf of St. Lawrence was ice-free. The difference between trends in arrival and departure represents average measured residency time. From Ramp et al., 2015, fig. 2.

by Ramp and colleagues (2015) revealed that the earlier arrival of fin and humpbacks—earlier by approximately one day per year over 27 years—was strongly related to ice break-up and rising sea surface temperature, indicating the whales' ability to adapt to a changing environment (figure 7.4).

An integrative study by Stanford biologist Elizabeth Alter and colleagues (2015) used genetic (ancient and mod-

ern DNA) and isotopic information to explore changes in the ecology of gray whales, *Eschrichtius robustus*, during the Pleistocene and Holocene. Their results indicate that dispersal of gray whales was climate-dependent and occurred during the Pleistocene prior to the last glacial period and the early Holocene, immediately following the opening of the Bering Strait. The genetic diversity of Atlantic gray whales declined over a period of time that

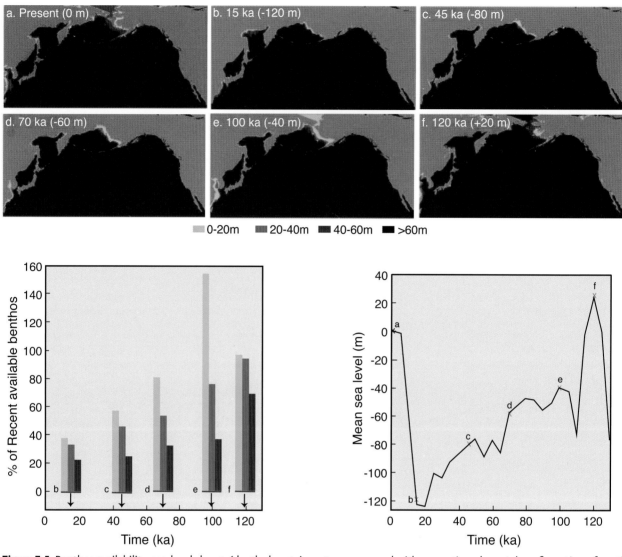

Figure 7.5. Benthos availability, sea-level change (depth change in meters compared with present), and coastal configuration of continental margins in the North Pacific Ocean at select intervals during the past 120,000 years. *Top*: Reconstructed coastal configurations and 20 m depth contours for (a) present day, (b) 15 ka, (c) 45 ka, (d) 70 ka, (e) 100 ka, and (f) 120 ka. *Bottom*: bar graphs with available benthos at 20 m increments at select time intervals (a–f). From Pyenson and Lindberg, 2011, fig. 2.

predates intensive commercial whaling, indicating that this decline may have been initiated by climate change or other ecological causes. Genetic data for present-day occurrences of Atlantic gray whales (off the coast of Israel and Spain) combined with habitat models suggest that these gray whale sightings may represent the beginning of an expansion of this species' habitat beyond its current North Pacific Ocean range.

In an earlier study, paleontologists Nick Pyenson and David Lindberg (2011) showed that changes in Pleistocene sea level cycles resulted in fluctuations in available benthic

feeding areas (figure 7.5). These scientists suggested that gray whales survived Pleistocene glacial maxima by feeding outside the Sea of Okhotsk and Bering region, which were ice-covered, employing more generalized feeding strategies. These strategies enabled them to feed on alternative prey such as fish, much like the feeding habit of seasonal resident gray whales found off the Pacific Northwest coast.

Whale molecular biologist Phil Morin and colleagues (2015) used mtDNA to show that killer whales, *Orcinus orca*, have undergone a rapid global diversification beginning in the middle Pleistocene, approximately 350,000 years ago, and that long-distance dispersal between hemispheres and ocean basins occurred throughout the late Pleistocene and Holocene (figure 7.6). These range expansions and contractions of killer whales were caused by glacial cycles. Habitat models show that only 15% contraction of core habitats occurred during the last gla-

cial maximum. One of the more interesting results of the study is that the rapid build-up of genetic differentiation and lineage sorting in killer whales led to ecological and geographic diversification.

Isotopic, genetic, and chronological data reported by a number of scientists, including Seth Newsome, Paul

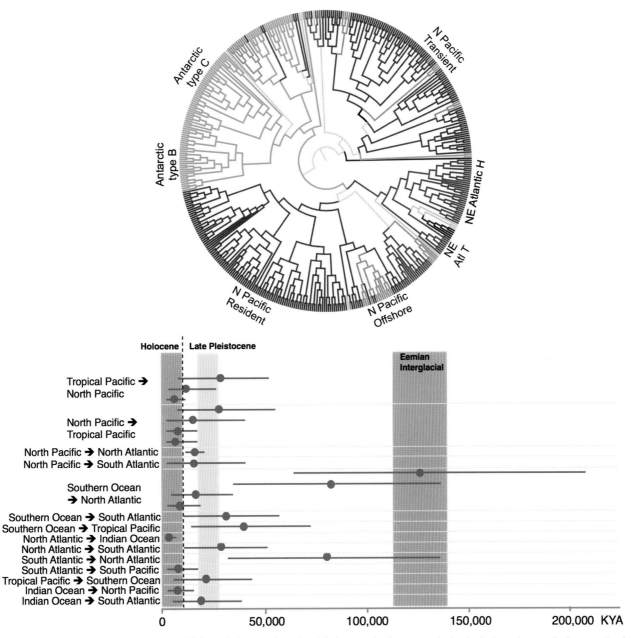

Figure 7.6. Cladogram (*top*) showing all killer whale samples (identified ecologically or morphologically based on well-characterized types or populations) in the study, with nodes and branches colored based on geographic location (see Morin et al., 2015, for detailed methods). *Bottom*, estimated dates of nodes at which the geographic location of killer whales is inferred to have changed from the ancestral range. *Orange bars* indicate the timing of interglacial periods; *blue bar* indicates the timing of the Last Glacial maximum. From Morin et al., 2015.

Koch, and Malin Pinsky, reveal prehistoric shifts in the ecology of northern fur seals, *Callorhinus ursinus*, from the Holocene to the present (e.g., Koch et al., 2009; Newsome et al., 2007, 2010; Pinsky et al., 2010). This species is common in archaeological middens from southern California to the Aleutian Islands. Today this species breeds almost exclusively on high-latitude offshore islands, with the largest rookeries in the Bering Sea (Pribilof Islands). Isotopic differences were found among seals from different regions, and archaeological evidence of young pups at sites in California, the Pacific Northwest, and the eastern Aleutians confirms the presence of temperate-latitude breeding colonies. In extant populations, a drop in nitrogen isotope values between age groups 2–6 months and 6–9 months is consistent with rapid weaning, when pups switch from feeding on maternal tissues to foraging independently (figure 7.7). In contrast, nitrogen isotope values from Holocene temperate latitudes indicate that *C. ursinus* pups were weaned at 12 months or older, not at 4 months as today. This species not only survived a dramatic collapse of its breeding range and heavy exploitation for the fur trade but also maintained high genetic variability when occupying sub-Arctic refugia (Pribilof Islands). High dispersal rates among ancient and current populations prevented loss of genetic diversity in this

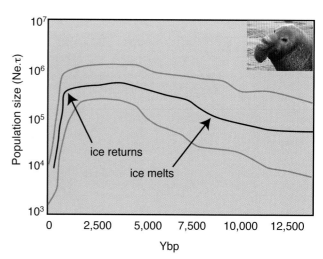

Figure 7.8. Bayesian skyline plot of population size of southern elephant seals. The *central black line* indicates the median posterior effective population size through time. *Gray lines* show the 95% highest posterior density interval, taking into account coalescent model and phylogenetic uncertainty. In this type of analysis, DNA sequence data from multiple individuals in a population are used to co-estimate genealogies and the effective population size at various points in time. *Abbreviation:* Ne.τ, generation time. From de Bruyn et al., 2011.

species, which suggests that panmictic species with large ranges will be more resilient to future disturbances and environmental change.

An aDNA study of southern elephant seals, *Mirounga leonina*, which have a circumpolar distribution, explored how these animals were affected by climate change from the late Pleistocene to today. Increased genetic diversity in southern elephant seals from the sub-Antarctic Macquarie Island and the Antarctic mainland was explained by rapid population growth as ice-free areas increased along the Ross Sea, about 7,000 years ago. The elephant seal population in the region contracted and lost genetic diversity beginning 1,000 years ago with the expansion of sea ice, well in advance of impacts from sealing in the early nineteenth century (figure 7.8). Perhaps most significantly, these data suggest that climate change could stimulate evolutionary change over a very short time scale. Genetic studies also reveal that historical population expansion of other Southern Ocean predators—notably Weddell seals, *Leptonychotes weddellii*; crabeater seals, *Lobodon carcinophaga*; and Ross seals, *Ommatophoca rossii*—

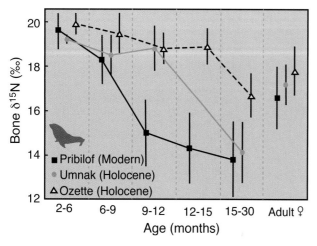

Figure 7.7. Nitrogen isotopic evidence for a large change in weaning age in northern fur seals from a modern rookery (Pribilof Islands) and Holocene populations from the Olympic Peninsula, Washington (Ozette), and the Aleutians (Umnak). From Newsome et al., 2010.

was most likely environmentally driven (Younger et al., 2016).

Recent studies show that the growth rates of marine mammal teeth are sensitive to climate change, and this is proposed as an underutilized tool for developing a historical record of climate-growth relationship. Study of the teeth of the New Zealand fur seal, *Arctocephalus forsteri*, indicates that mean annual sea surface temperature is the most important climate factor influencing interannual tooth growth. Based on these results, Wittemann and colleagues (2016) propose that marine mammal teeth provide critical information about past and future ecological responses to environmental change, which is worthy of further exploration in other marine mammal species.

Habitat Loss

The loss of sea ice poses a threat to Arctic marine mammals such as the walrus, *Odobenus rosmarus*, the polar bear, *Ursus maritimus*, and the hooded seal, *Cystophora cristata*, which rely on sea ice as a platform for resting and reproduction. Observations of walrus calves swimming away from shore suggest that mothers are forced to abandon pups with the retreating ice. Polar bears may drown because of the increasing travel distances between receding ice sheets. Changes in ice thickness and extent also affect coastal habitat and the species that rely on it. The decline in sea ice has reduced the body size of polar bears and decreased their reproductive output and juvenile survival (Molnar et al., 2011). The warming of the ocean has changed species ranges, including migration routes and timing of migrations. For example, scientists have observed in recent years that gray whales delay their southward migration to the breeding lagoons off Baja California, Mexico. As warmer waters melt the ocean's ice, other animals move into the whales' habitat and start feeding on the crustaceans. Crowded out by the new competition, the gray whales have to travel further north and feed longer to get their fill. These changes have disrupted the timing of the gray whales' yearly migration.

Using global climate models, scientists can predict increases in temperature and receding ice. Temperature increases have resulted in the disappearance of krill in polar regions, with some estimates indicating as much as an 80% loss of krill in the Antarctic in the past 40 years. The loss of sea ice removes a primary source of food for krill—the algae that grow on ice. Given the significance of krill to the diets of many marine species, including crabeater seals and most baleen whales, its disappearance is alarming and cause for concern. Changes in temperature will also change the location of areas with high primary productivity. These areas are important to marine mammals because primary producers are the food source for marine mammals' prey. For example, while the numbers of blue whales, *Balaenoptera musculus*, feeding off the California coast have decreased in the past 10 years, sightings of these whales have increased in the northern waters off Canada and Alaska. Research suggests that blue whales may be migrating north in search of krill, which has been depleted in the waters off California due to climate warming.

Changing Food Web Dynamics

Food webs form the framework of marine mammal interactions because they define (or at least describe) how energy flows through the ecosystem from primary producers to various consumers. Nearly all food webs are based on photosynthesizers, although a few marine food chains (e.g., whale fall communities; see below) depend on bacterial chemosynthesizers, which use chemical sources of energy for production. Sirenians are unique among marine mammals in feeding at the base of the food chain on sea grasses and marine algae. All other marine mammals are carnivores (mostly piscivores) that consume both herbivores and other meat/fish-eaters at higher trophic levels. The food web dynamics of ecosystems are regulated by "top-down" and "bottom-up" processes (figure 7.9). "Top down" refers to control exerted within the biological system that stems from top predators (predators controlling prey densities). "Bottom up" refers to control exerted upward in the food chain, driven by the physical properties (e.g., climate) that determine the availability of prey "resources" in the system (see Ainley et al., 2007). Sea otters, polar bears, and killer whales are all top predators in marine food chains, acting

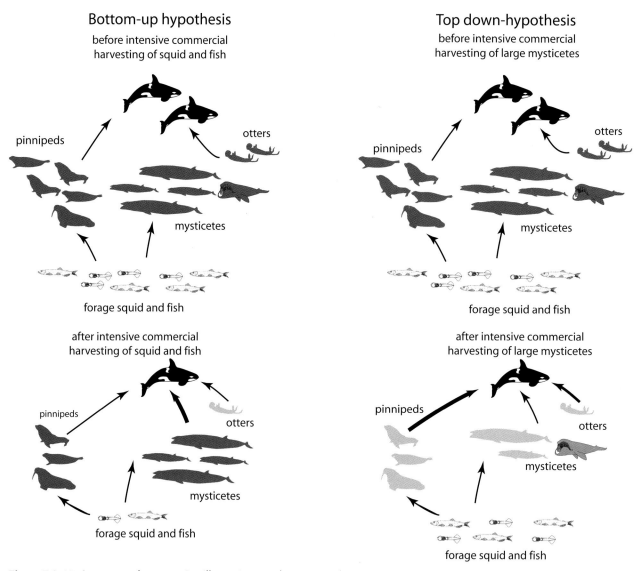

Figure 7.9. Marine mammal community, illustrating top-down versus bottom-up processes. *Arrow thickness* indicates relative intensity of foraging by killer whales on two different prey types.

as dominant predators or keystone species that control species abundance and diversity at lower levels. For example, climate-mediated, top-down effects of polar bears in the Arctic have a positive impact (reduced predation) on Arctic seals (e.g., ringed seal, *Pusa hispida*) and negative impact (increased predation) on sub-Arctic seals (e.g., harp seal, *Pagophilus groenlandicus*), nesting seabirds (eiders), and terrestrial resources (figure 7.10).

SEA OTTERS AS KEYSTONE SPECIES

Sea otters, *Enhydra lutris*, in the North Pacific Ocean are considered keystone species, playing a pivotal role in

kelp bed communities by keeping populations of sea urchins (which feed on kelp) low (figure 7.11). The Australasian biotic assemblage in the southern hemisphere, in contrast, lacks a predator comparable to the sea otter. In such a food web, kelp is left unprotected. Instead, chemical defenses have evolved to protect against herbivory.

INTERACTIONS AMONG SEA OTTERS, KELP FORESTS, AND STELLER'S SEA COW

Close relatives of the extinct Steller's sea cow, *Hydrodamalis gigas*, had a wider distribution in the late Pliocene and Pleistocene, extending from Japan to the Aleutians

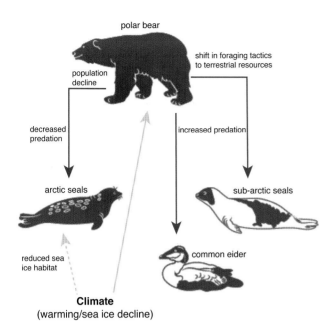

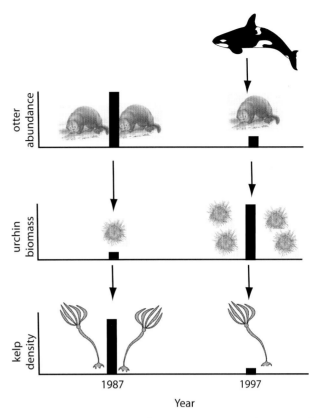

Figure 7.10. Climate-mediated top-down effects of polar bears in the Arctic, with positive and negative impacts (see text). *Orange arrows* indicate direct climate controls; *red arrows*, top-down interactions. *Solid lines* indicate well-supported effects; *dashed lines*, hypothesized effects. From Sydeman et al., 2015.

Figure 7.11. Comparison of ecosystem in coastal waters of Aleutian Islands in 1987 and in 1997, when sea otters declined, sea urchins increased, and kelp decreased. *Vertical bars* are relative indices of the variable listed. Note that the 1997 sequential collapse is hypothesized to have been caused by killer whales "fishing down" the food web. Adapted from Estes et al., 1998.

to the Pacific coast of Baja California. Steller's sea cows disappeared from most of their range between the Pleistocene and the arrival of Europeans in the North Pacific in the eighteenth century. The Commander Islands were a notable exception to this widespread extinction, and an abundance of Steller's sea cows survived there in 1741 when the Bering Expedition shipwrecked and overwintered on the islands. Whether the range retraction of these mammals was due to cooling temperatures that limited them to the Commander Islands, perhaps the indirect effect of reduced food, or was due to hunting by native peoples is still debated (see also chapter 5). A third possibility is that their extinction occurred through habitat changes resulting from the loss of other keystone species. This was the finding of a study by biologist Jim Estes and colleagues (2015), which suggested that extinction of Steller's sea cow was a consequence of the loss of sea otters and the co-occurring loss of kelp (figure 7.12). Evidence came from modeling of the sea cow's response by applying known demographic and behavioral responses of dugongs (the sea cow's closest living relative) to cat-

astrophic food loss. Simulations run by these scientists showed that sea cows around Bering Island (their supposed last population) would have reached near or complete extinction by or close to 1768, the year of their last reported sighting. Although human hunting may have been largely responsible for the decline of *H. gigas* at Bering Island, its extinction was also a nearly inevitable consequence of the loss of sea otters and kelp forests and would have occurred without a single human take of the species.

PREDATION AND TROPHIC DOWNGRADING

Predation by killer whales is thought to be one of the major factors affecting populations of many marine mammal species. The degree to which the killer whale preyed upon certain marine mammals during the onset of their

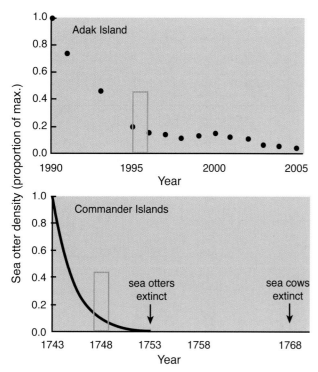

Figure 7.12. Trajectories of sea otter declines in the Aleutian Islands during the 1990s and early 2000s (*top*) and in the Commander Islands after onset of the fur trade in 1743 (*bottom*). The *line in the bottom graph* assumes that sea otters were at maximum density in 1743 and extinct by 1753 and that the decline was exponential. *Open blue boxes* indicate the time window of kelp forest phase shift at Adak Island (*top*) and the estimated time of this phase shift in the Commander Islands. From Estes et al., 2015.

decline is unclear, however, but recent evidence suggests that this may be a factor currently preventing such species from recovering from their depleted status. To further complicate the issue, killer whales are declining in numbers where the Steller sea lion, *Eumetopias jubatus*, and sea otter populations are in crisis but are abundant where Steller sea lions are showing signs of recovery.

An extension of the predation hypothesis is that after peak commercial whaling, the killer whale no longer had the large baleen whales (fin whales, sei whales) and sperm whales to prey upon as their populations were decimated. This prompted a "top-down" shift in predation to Steller sea lions in the early 1970s in the Gulf of Alaska, Aleutian Islands, and Bering Sea. The population decline of Steller sea lions and the slow recovery rate of cetacean popula-

tions caused killer whales to switch again, sequentially, to the other marine mammal populations that have recently declined: harbor seals, northern fur seals, and, finally, sea otters. Evidence to support this "sequential megafaunal collapse" hypothesis is the timing of the collapse and the known diets and observed foraging behavior of killer whales and their prey (Springer et al., 2003, 2008). The removal of large apex consumers has been referred to as "trophic downgrading" and is recognized in marine, terrestrial, and freshwater systems worldwide (Estes et al., 2011). An alternative "bottom-up" model proposes that reduced nutrient supply and productivity are the culprits, which in the case of Steller sea lions and other prey suggests that they have collapsed because their prey population has declined (DeMaster et al., 2006; see also the discussion below and figure 7.9). In another study of ecosystem dynamics, this time in the Southern Ocean, some scientists proposed a top-down explanation, also referred to as the "krill surplus hypothesis." This hypothesis suggests that removal of large whales led to a krill surplus that resulted in expansion of other predator populations such as penguins and minke whales, since they were no longer competing with large whales. Others, including NOAA scientist Lisa Ballance and colleagues (2006), proposed that this system is driven by "bottom-up control" of the abundance of krill in the system, with variation in krill recruitment and density correlated with ice cover. Although there has been much interest and lively debate about these hypotheses (e.g., Ruegg et al., 2010), clearly, both top-down and bottom-up processes are important at present and most likely in the past. Disentangling the causative effects of ecosystem change requires, for a particular spatial scale (i.e., region), the development of rigorous hypotheses that incorporate biological as well as physical drivers of ecosystem change.

Lindberg and Pyenson (2006) examined Springer's top-down hypothesis with a different twist, considering body size rather than trophic level in the context of the fossil record. Their results suggested that ancestral cetaceans in major lineages (sperm whales, delphinids, beaked whales), by increasing their body size, successively "fished up" prey size—that is, fed on larger prey through evolutionary time. The fishing up of prey size probably began with "archaeocetes," or perhaps even earlier with marine reptiles. In addition, cetaceans evolved to take relatively small but abundant prey, which probably selected for an increase in cetacean size given the increase in prey size. Among delphinids, the fossil record of the killer whale indicates that ancestral forms were smaller (approximately 4 m [13 ft] long) than extant species but larger than most delphinids. This is consistent with the evolutionary fishing-up hypothesis suggesting the evolution in extant killer whales of the ability to feed on large whales. This pattern of larger delphinid ancestors that fished up to larger prey items is also seen in late Miocene ziphiids and physeterids and the mysticetes. The decline of mysticetes during the Pliocene coincides with the appearance and diversification of their likely predators: carcharodontid sharks (see chapter 3). Although these predators fed on mysticetes during the Neogene, determining from the fossil record whether predation was directly responsible for the extinction of some mysticete lineages presents a challenge.

Whale Falls and Fossil Whale Fall Communities

Like the animal communities around deep-sea hydrothermal vents, whale fall communities evolve around food sources. Most are found in deep-water settings. At hydrothermal vents, sulfides spewing from the vents are consumed by bacteria, which in turn provide nutrition for animals. Similar food webs have been observed at whale falls, where sulfides may be produced by bacterial decay of whale tissue. The sulfophilic (sulfur-loving) whale fall community differs from typical food webs in that, instead of being based on photosynthetic organisms, it is based on chemosynthetic bacteria. Roughly 10% to 20% of sulfophilic species found at whale falls are also found in hydrothermal vent communities, but the majority of species found in the two types of environment are unique to each.

Like marine biologists, paleontologists have been interested in these deep-sea whale fall communities. There is evidence that whale falls remain and decompose on the seafloor as well as at relatively shallow depths. They

pass through a series of overlapping successional stages that vary on the basis of carcass size, water depth, and environmental conditions. Whale falls have served as hot spots of adaptive radiation for a specialized fauna, and they could have facilitated speciation in other vent/seep taxa. Fossil whale fall communities have been discovered at more than 50 sites in North America, Europe, Asia, and Africa (Smith et al., 2015).

In the first, mobile scavenger stage of succession, sharks and hard-shelled invertebrates (e.g. whelks), associated with most fossil whale falls, are responsible for removal of soft tissue. The subsequent enrichment opportunist stage is dominated by deposit-feeding pelecypods. Bacterial grazers such as some gastropods have been found on whale skeletons from the Miocene to the present, and bone-eating *Osedax* worms have been found in whale bones from the Oligocene to the present. The sulfophilic stage is represented by naticid and scaphopod molluscs that colonize the bones and sediments. The final stage is colonization primarily by suspension-feeders such as oysters and barnacles, which are common in shallow-water whale falls (figure 7.13). Other taphonomic studies (e.g., Danise and Domenici, 2014) found that shallow-water whale falls may not go through all the stages of ecological succession (the sulfophilic stage is not well represented in the fossil record).

Molecular studies suggest that whale falls have served as hot spots of speciation and evolutionary novelty and may have served as evolutionary or ecological "stepping stones" for vent/seep faunas. The fossil record is consistent with this interpretation.

Strandings

Dead marine mammals strand and wash ashore around the world. Strandings have various causes, including disease, pollution, military sonar, and changes in prey resources linked to cyclic climatic events. By compiling and comparing stranding records, Nick Pyenson (2011) demonstrated that strandings (death assemblages) of cetaceans realistically record ecological snapshots of living communities, including number and diversity of species. The results of Pyenson's study suggest that some fossil

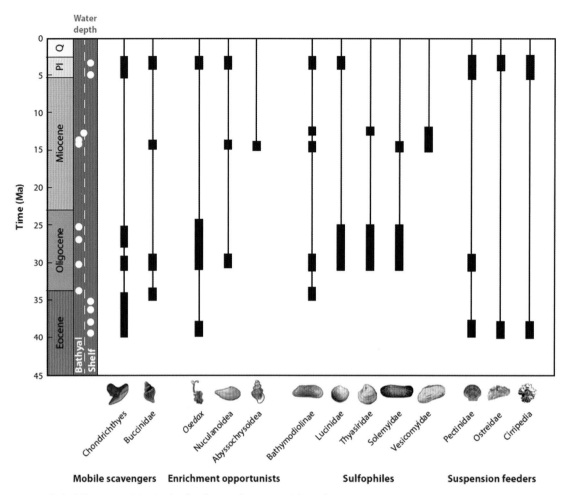

Figure 7.13. Whale fall communities in the fossil record. From Smith et al., 2015.

assemblages might similarly record paleoecological data, which could provide clues to the structure and abundance of cetacean communities.

Mass strandings of marine mammals, which can involve hundreds of individuals, are mostly restricted to odontocetes. Such strandings are recorded in historical times, but discovering a causal mechanism for fossil mass strandings is challenging. Harmful algal blooms (HABs) caused by species of phytoplankton, some of which produce potent chemical toxins, have been implicated in some mass stranding events. Fueled by periodic abundances of nutrients in the ocean, these algae multiply and proliferate until they can cover tens to hundreds of miles of coastal ocean. HABs that release domoic acid have re-

sulted in mass mortalities of marine mammals. One death assemblage of fossil marine mammals included 40 skeletons of balaenopterids, sperm whales, seals, and aquatic sloths from the late Miocene Atacama region of Chile (figure 7.14). This assemblage was explained by Nick Pyenson and colleagues (2014) as being caused by HABs through ingestion of prey or inhalation, causing relatively rapid death at sea. Carcasses were thought to have floated toward the coastline in large numbers and hence were buried without major scavenging or disarticulation. The conditions that lead to HAB-mediated mass strandings are tied to upwelling systems along continental margins, consistent with the hypothesized occurrence along the western South American coastline.

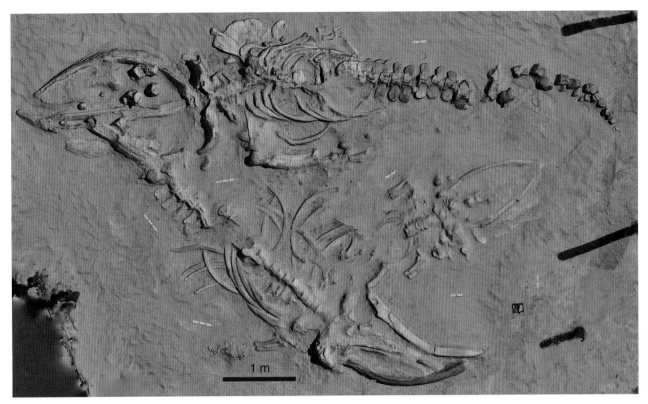

Figure 7.14. Overlapping adult and juvenile rorqual specimens at Cerro Ballena, Chile. From Pyenson et al., 2014.

Disease

Evidence for the effect of disease on fossil marine mammals most often involves bone pathologies. Osteochondrosis, a disease that affects bone joints, has been reported in various extant and extinct marine mammals, including pinnipeds, cetaceans, and desmostylians. The pathology is indicated by scars on the articular surfaces of bones (e.g., shoulder, elbow, and wrist joints). Based on these occurrences, the injuries may have been caused by interactions with other animals or, in the case of pinnipeds and desmostylians, when the animal was entering or leaving the water (Thomas and Barnes, 2015).

Pollutants and Stress

Marine mammals respond to the prevalence and persistence of pollutants in marine ecosystems by increasing their levels of stress hormones. For example, Baylor

University scientist Stephen Trumble and his colleagues (2013) found that ear plugs in the blue whale, *Balaenoptera musculus*, recorded contaminants (e.g., mercury, pesticides) and hormones (e.g., cortisol, testosterone) that routinely accumulate in whale blubber. The contaminant and hormone levels measured in the waxy layers of ear plugs can be used to reconstruct lifetime profiles of contaminant exposure and stress in individual whales. This is likely to be true for other mysticete species, too, since they also possess ear plugs.

A record of increased levels of stress and reproductive hormones (cortisol, progesterone) was found in the baleen of bowhead whales, *Balaena mysticetus*. These data suggest that baleen may be a useful new tool for reconstructing the past life of baleen whales, which is likely to prove valuable in reconstructing patterns of stress and reproductive cycles. For example, changes in calving intervals, whether caused by climate change or increases in human activity, could allow the tracking of long-term population changes and life history patterns through time (Hunt et al., 2014).

Extinctions and Interactions between Humans and Marine Mammals
Extinction: The Rule, Not the Exception

Extinction is a normal process. Species have a natural duration, from a few thousand to a few million years, so they live for a period of time and then disappear—or become extinct. The most spectacular extinctions are mass extinctions, when a large number of different species die out relatively rapidly. At least five mass extinctions have been identified in the geologic past but only one has affected marine mammals, given their relatively recent evolutionary history. The current extinction crisis, which began in the late Pleistocene, is often described as the sixth mass extinction event, the only one that is human-caused. Humans have accelerated the rate of extinction of many marine species, including marine mammals, through overfishing and overhunting, habitat loss and degradation, pollution, and global warming.

Large marine vertebrates, including whales, sea cows, and monk seals, are now functionally or entirely extinct in most coastal ecosystems. A compelling case has been made that overfishing is the critical driver of human disturbance in ecosystems. For example, historical records document the extensive hunting of manatees, *Trichechus* spp., by early colonists in the Americas and of dugongs by aboriginal people of Australia. This overfishing increased the vulnerability of sea-grass beds—which provide food and habitat for manatees and dugongs, among other vertebrates—to recent events such as increases in sedimentation, turbidity, and disease. The disruption of sea-grass ecosystems ultimately led to increasingly fragmented populations of once large herds of dugongs, *Dugong dugon,* and manatees (Jackson et al., 2001). This and other examples of changing ecosystem dynamics (e.g., killer whales and sea otters, discussed above) point up the importance of collecting historical data that document human exploitation of coastal resources for food and mate-

rials (see Braje and Rick, 2011). These historical data serve as an important management tool for ecosystem restoration because they reveal that recent events (e.g., climate change, pollution) are often symptoms of problems with deep historical causes.

Can Biologists Predict Future Extinction Rates?

The rate of extinction of plant and animal species in the past 500 years is thought to be 1,000–10,000 times higher than it would have been without the effect of humans. Assessing the probability of extinction of marine mammals is difficult, because it necessitates determining the size of wild populations and changes in populations during a specified time interval. There are also uncertainties in the relative importance of different drivers of extinction and potential interactions among drivers, a lack of basic assessment for many species, and spatial uncertainty in climate projections. The International Union for the Conservation of Nature (IUCN) identifies 25% of marine mammals as at risk of extinction, but the conservation status of nearly 40% of marine mammal is unknown due to insufficient data. In a study by biologist Ana Davidson and colleagues (2012) that quantified extinction risk for marine mammals, the primary predictors of risk were two intrinsic factors: body mass at weaning and number of births per year (figure 7.15). These two life history variables inform the rate of reproduction, and high-risk species such as sirenians are species with low rates of reproduction and slow life histories. Other important predictors are small geographic range and small social-group size; the latter is a factor because of the advantages of sociality in reducing predation. Among the leading extrinsic threats to marine mammals worldwide found in Davidson et al.'s study were overfishing and bycatch. Shipping and pollution affect marine mammals through ship strikes and noise and other forms of pollution such as oil spills, discarded fishing gear, and ingested plastic debris. These effects are widespread throughout the northern hemisphere, especially along the eastern and western North Pacific coasts, which are major population centers and shipping routes.

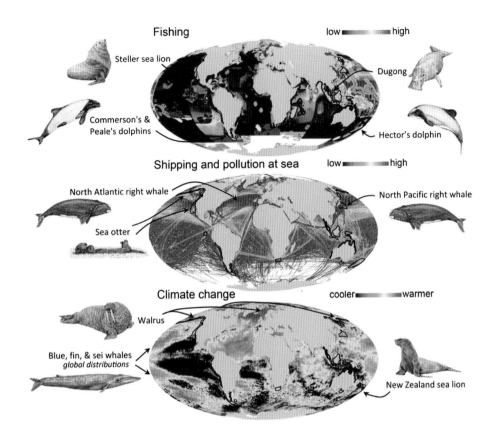

Figure 7.15. Global hot spots of marine mammal species, overlaid with geographic distribution of the leading human impacts. From Davidson et al., 2012.

Extinction rates differ significantly among major lineages of marine animals (figure 7.16). The rates among marine mammals (cetaceans, marine carnivorans, sirenians) are more than 10 times those of most invertebrates.

Knowledge of past extinction patterns is critical for predicting future extinction vulnerability. A study of extinction risk among marine taxa (including pinnipeds and cetaceans), determined by examining baseline data analyses of extinctions over the past 23 million years (Miocene-Pleistocene), revealed that geographic range and taxonomic identity are among the most consistent predictors of extinction risk in the marine fossil record (Finnegan et al., 2015). Not surprisingly, both factors were also identified in the Davidson et al. (2012) study of extinction risk based on modern global biodiversity patterns.

The first well-documented human-caused extinction in historical times was that of Steller's sea cow, *Hydrodamalis gigas*. Humans are also responsible for the extinction of the Chinese river dolphin, *Lipotes vexillifer*, de-

clared extinct less than 10 years ago as a result of habitat loss, and the Caribbean monk seal, *Neomonachus tropicalis*, hunted to extinction in the twentieth century. Several other marine mammal species hover on the brink of extinction, such as the vaquita, *Phocoena sinus*, the victim of bycatch in gill-net fishing in the Gulf of California, and the North Atlantic right whale, *Eubalaena glacialis*, with mortalities largely caused by ship collisions.

Extinction and Replacement: Human-Mediated Shifts

In an example of extinction and replacement that were human-caused, Collins et al. (2014) used aDNA analysis to show the genetic distinctiveness of subfossil populations of the endemic New Zealand sea lion, *Phocarctos* spp., of mainland New Zealand compared with modern populations. The extinction of *Phocarctos* spp. caused by humans soon after their arrival in AD 1280–1450 apparently facilitated a northward range expansion of *Phocarctos* popula-

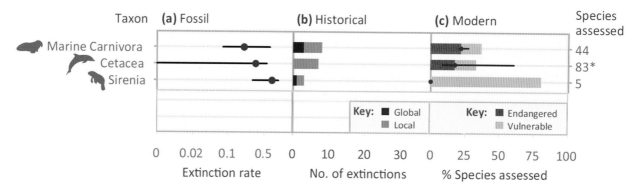

Figure 7.16. Comparison of (*a*) extinction rates in the Cenozoic fossil record, (*b*) historical record of number of extinctions, and (*c*) current risk of extinction among marine carnivores, cetaceans, and sirenians. (*a*) Median (*circles*) and 1st and 3rd quartiles (*lines*); (*b*) number of global extinctions (*dark gray*) and local extirpations (*light gray*); (*c*) percentage of modern species assessed by the IUCN as endangered (*dark gray*) and vulnerable (*light gray*), excluding data-deficient species. Numbers on the right indicate the number of modern species assessed by the IUCN for each taxonomic group. Asterisk indicates taxonomic group with >50% of assessed species considered data-deficient. Line segments in (*c*) indicate upper and lower estimates of the fraction of endangered species if all data-deficient species were classified as endangered or not endangered, respectively. From Harnik et al., 2012.

tions in mainland New Zealand, thereby replacing a previously sub-Antarctic-limited lineage.

Final Thoughts

The returns to the sea of the various marine mammal lineages from their land mammal ancestries represent major macroevolutionary transitions in ocean ecosystems. Data from paleontology, anatomy, molecular biology, and developmental biology help trace this transition from land to water. The transition occurred independently at least seven times among marine mammal lineages, including whales and sirenians during the Eocene (50 million years ago), pinnipeds during the late Oligocene (30 million years ago), desmostylians during the late Oligocene to early Miocene (33-10 million years ago), extinct aquatic sloths during the late Miocene and early Pliocene (7.8-1.5 million years ago), and the sea otter and polar bear since the Pleistocene (1 million years ago). The morphological and physiological adaptations that resulted from these habitat shifts enabled marine mammals to efficiently locate and process aquatic food sources, adopt deep div-

ing, and employ limb- and tail-based propulsion in the water. Marine mammals have been extraordinarily successful over the past 50 million years, and they remain ecologically important in the twenty-first century. They have occupied apex roles in food webs, feeding both at the base of the food chain and at higher trophic levels throughout major changes in oceans and climate. Understanding how drivers of past diversity, both environmental and biotic, continue to operate today helps shape our efforts to conserve and manage these remarkable mammals of the sea.

There is growing recognition that an awareness of the processes that occur over evolutionary timescales is important for biodiversity conservation (Sarrazin and Lecomte, 2016). For it is fossils that provide the critical, unique reference points for understanding historical changes in marine mammal communities and diversity through time. The fossil record thus provides a vital context for evaluating the ongoing biodiversity crisis that is affecting marine mammal populations and global marine ecosystems.

Classification of Fossil Marine Mammals

An important reference is the list of the taxonomy of extant marine mammal species and subspecies provided by the Committee on Taxonomy of the Society of Marine Mammalogy (2016). In the following classification, the most significant changes to traditional higher taxonomic arrangements are provided under "Remarks." Emphasis is placed on providing content (included genera) and distributional and fossil record data for monophyletic groups at the "family" level. The definition of a taxon is based on ancestry and taxonomic membership, with a focus on well-documented fossil taxa discussed in this book. For the most part, higher-level classification follows the Paleobiology Database (www.pbdb.org). For pinnipeds, see Berta et al. (1989); cetaceans, Geisler et al. (2011); and sirenians, Vélez-Juarbe et al. (2012). See the text for more detail on taxonomic membership. For more information on geographic distributions of extant taxa, see data in the IUCN Red List (IUCN, 2015) and Jefferson et al. (2015). Quotation marks are used for taxa for which the monophyly is questioned, and a dagger designates extinct taxa. The timescale used (see chapter 1, figure 1.18) follows Gradstein et al. (2012). Ma = million years ago; kya = thousand years ago. Temporal and age distributions for fossil taxa follow the Paleobiology Database.

CARNIVORA Bowditch 1821
   ARCTOIDEA Flower 1869
      PINNIPEDIMORPHA Berta et al. 1989
The earlier-diverging pinnipedimorphs *Enaliarctos*† and *Pteronarctos*† are known from the late Oligocene through the early Miocene (27-16 Ma), North Pacific (North America, Japan) (see Deméré et al., 2003). Fossil pinnipedimorphs are reviewed by Berta (2009).
         Enaliarctos† Mitchell and Tedford 1973
         Pacificotaria† Barnes 1992
         Pinnarctidion† Barnes 1979
         Pteronarctos† Barnes 1989
        PINNIPEDIA Illiger 1811

Content—Three families of living pinnipeds are known. A number of fossil genera are included (see below), among which are several not belonging to modern families (e.g., Desmatophocidae†).

Distribution—Pinnipeds occur in all of the world's oceans.

Fossil history—Pinnipeds are recognized in the fossil record beginning in the middle Miocene (20 Ma)-Pleistocene, North Pacific (North America, Japan); late Miocene-Pleistocene, eastern South Pacific (South America); late middle Miocene-Pleistocene, North Atlantic (western Europe, North America); latest Miocene-early Pliocene, western South Pacific (Australasia); early Pliocene, eastern South Atlantic (South Africa) (see Deméré et al., 2003).
         OTARIIDAE Gill 1866

Content—Extant genera comprise 7 genera with 15 species. Taxonomy follows the Committee on Taxonomy (2016) and Berta and Churchill (2012).

Distribution—Otariids are distributed throughout the world except for the extreme polar regions.

Fossil history—Extant genera with a fossil record include *Neophoca* Gray 1866, early Pleistocene, South Pacific (New Zealand); *Arctocephalus* Cuvier 1826, Pliocene (5-2.7 Ma), late Pleistocene (330-1 kya), South Pacific (Africa), and late Pleistocene, South America; *Callorhinus* Gray 1859, late Pliocene-Pleistocene (3.6 Ma-Recent), eastern North Pacific (California, Japan, Mexico); *Eumetopias* Gill 1866, Pleistocene (2.588 Ma-Recent), eastern North Pacific (Japan) and eastern South Pacific (South America).

Otariids have a fossil record beginning in the early middle Miocene (20.43-13.65 Ma)-Pleistocene, eastern North Pacific; Pleistocene, eastern South Pacific (Peru). Extinct genera are reviewed by Churchill et al. (2014).

> *Eotaria*† Boessenecker and Churchill 2015
> *Hydrarctos*† Muizon 1978
> *Oriensarctos*† Mitchell 1968
> *Pithanotaria*† Kellogg 1925
> *Proterozetes*† Barnes et al. 2006
> *Thalassoleon*† Repenning and Tedford 1977

ODOBENIDAE Allen 1880

Content—One genus and species, *Odobenus rosmarus* Linnaeus 1758, and two subspecies (*O. rosmarus rosmarus* Pallas 1861, Atlantic walrus, and *O. r. divergens* Illiger 1815, Pacific walrus) are recognized. A third purported subspecies, *O. r. laptevi* Chapskii 1940 from the Laptev Sea, was found not valid by Lindqvist et al. (2009), who concluded that it should be recognized as the westernmost population of the Pacific walrus.

Distribution--Extant walruses inhabit the shallow, circumpolar Arctic coasts.

Fossil history—The extant genus *Odobenus* Brisson 1762 has a fossil record that extends back to the Pliocene (2.588 Ma). Fossil walruses are known as far back as the early middle Miocene (16 Ma)-Pleistocene, eastern North Pacific (North America); early Pliocene-Pleistocene, North Atlantic (western Europe); late Pleistocene, eastern North Pacific (Japan) (see Deméré et al., 2003). Extinct taxa are reviewed by Boessenecker and Churchill (2013).

> *Archaeodobenus*† Tanaka and Kohno 2015
> *Imagotaria*† Mitchell 1968
> *Kamtschatarctos*† Dubrovo 1981
> *Neotherium*† Kellogg 1931
> *Pelagiarctos*† Barnes 1988
> *Pontolis*† True 1905
> *Proneotherium*† Kohno et al. 1995
> *Prototaria*† Takeyama and Ozawa 1984
> *Pseudotaria*† Kohno 2006
> > **Dusignathinae** Mitchell 1968
> > *Dusignathus*† Kellogg 1927
> > *Gomphotaria*† Barnes and Raschke 1991
> > **Odobeninae** Mitchell 1968
> > *Aivukus*† Repenning and Tedford 1977
> > *Ontocetus*† Leidy 1859
> > *Protodobenus*† Horikawa 1995
> > *Valenictus*† Mitchell 1961

PHOCIDAE Gray 1821

Content—Extant phocids comprise 14 genera and 19 species. Taxonomy follows Berta and Churchill (2012) and Scheel et al. (2014), as modified by the Committee on Taxonomy (2016).

Distribution—Phocids are distributed throughout the world.

Fossil history—Extant genera with a fossil record include *Phoca* Linnaeus 1758, Miocene (12.7 Ma) of Europe (Austria), Pliocene of the United Kingdom, United States, and Quaternary of Canada, Europe, United States; *Mirounga* Gray 1827, middle late Pleistocene of the South Atlantic (Chile), late Pleistocene of South Africa, early Pleistocene of North Pacific (California), with questionable records from the Miocene (20.43 Ma) of South Africa and early Pleistocene of New Zealand; *Lobodon* Gray 1844, *Hydrurga* Gistel 1848, late Pleistocene-Holocene of South Africa; *Ommatophoca* Gray 1844, early Pleistocene (or late Pliocene) of New Zealand; *Erignathus* Gill 1866, Quaternary (0.126 Ma)-Recent of Canada, Europe, United States; *Pagophilus* Gray 1844, Pleistocene (1.806 Ma)-Recent of Canada, Europe, United States; *Pusa* Scopoli 1777, late Pleistocene-Recent of Canada.

A record of phocids from the late Oligocene (29-23 Ma) has been reported by Koretsky and Sanders (2002). However, the stratigraphic provenance of the specimen is uncertain. Phocids are well documented from the middle Miocene-Pleistocene, western North Atlantic (western Europe), eastern North Atlantic (Maryland, Virginia); early Pliocene-Pleistocene, South Pacific (Africa); middle Miocene-Pleistocene, South America. Extinct genera are reviewed by Berta (2009), Amson and Muizon (2013), Koretsky and Domning (2014), and Valenzuela-Toro et al. (2016).

   Devinophoca† Koretsky and Holec 2002

  Monachinae Trouessart 1897

   Acrophoca† Muizon 1981

   Afrophoca† Koretsky and Domning 2014

   Australophoca† Valenzuela-Toro et al. 2016

   Callophoca† Van Beneden 1876

   Hadrokirus† Amson and Muizon 2013

   Homiphoca† Muizon and Hendey 1980

   Monotherium† Van Beneden 1876

   Palmidophoca† Ginsburg and Janvier 1975

   Piscophoca† Muizon 1981

   Pliophoca† Tavani 1941

   Pontophoca† Kretzoi 1941

   Pristophoca† Gervais 1853

   Properiptychus† Ameghino 1887

  Phocinae Gray 1821

   Batavipusa† Koretsky and Peters 2008

   Cryptophoca† Koretsky and Ray 1994

   Gryphoca† Van Beneden 1876

   Kawas† Muizon and Bond 1982

   Leptophoca† True 1906

   Monachopsis† Kretzoi 1941

   Necromites† Bogachev 1940

   Phocanella† Van Beneden 187

   Piscophoca† Muizon 1981

   Platyphoca† Van Beneden 1876

   Praepusa† Kretzoi 1941

   Prophoca† Van Beneden 1877

DESMATOPHOCIDAE† Hay 1930

Fossil History—Desmatophocids are known from the early middle Miocene (23-13 Ma) of the North Pacific.

Allodesmus† Kellogg 1922

Atopotarus† Downs 1956

Brachyallodesmus† Barnes and Hirota 1995

Desmatophoca† Condon 1906

Megagomphos† Barnes and Hirota 1995

MUSTELIDAE Fischer 1817

Content—Extant mustelids comprise 22 genera and 50+ species, including the sea otter (*Enhydra*) and river otters (*Aonyx, Hydrictis, Lontra, Lutra, Lutrogale, Pteronura*).

Distribution—Mustelids occur worldwide except in Australia and Antarctica.

Fossil History—The extant sea otter *Enhydra* Fleming 1828 has a fossil record that extends from the Pliocene (2.588 Ma) to the present. Fossil sea otters have a record that extends to the Miocene (13.65-3.6 Ma) of the North Pacific. The extinct sea mink *Neovison* Baryshnikov and Abramov 1997 is recognized as a distinct species (see Mead et al., 2000) that formerly occurred along the coasts of Canada (New Brunswick, Newfoundland) and in coastal eastern North America (Massachusetts, Maine).

Enhydritherium† Berta and Morgan 1985

Neovison† Prentis 1903

AMPHICYNODONTIDAE† (Simpson) 1945)

Fossil History—Fossil taxa in the paraphyletic family Amphicynodontidae include *Amphicynodon, Pachycynodon, Allocyon,* and *Kolponomos*. The extinct bear-like carnivoran *Kolponomos* is known only from the Miocene (20.4-13.6 Ma) of Oregon, Washington, and possibly Alaska.

Kolponomos† Stirton 1960

URSIDAE Gray 1825

Content—Extant bears include 5 extant genera and 7 species.

Distribution—Ursids occur mostly in the northern hemisphere (Asia, North America, Europe), with the exception of the spectacled bear (*Tremarctos*), which occurs in South America.

Fossil History—The extant polar bear *Ursus maritimus* Phipps 1774 has a fossil record that extends into the Quaternary (0.012 Ma-present).

XENARTHRA Cope 1889

MEGATHERIIDAE Gray 1821

Fossil History—Nothrotheriids are extinct sloths that comprise 10 genera that lived in North and South America. A single genus of aquatic sloth is known, *Thalassocnus* from the Miocene-early Pliocene (7.2-5.88 Ma) of Peru and Chile.

Thalassocnus† Muizon and McDonald 1995

CETARTIODACTYLA Montgelard et al. 1997

CETACEA Brisson 1762

Content—Cetaceans comprise two clades (usually referred to as suborders), Mysticeti and Odontoceti, containing all living cetaceans.

Distribution—Cetaceans are distributed throughout the world's oceans.

Fossil history—Cetaceans are recognized in the fossil record beginning in the early Eocene (about 54-53 Ma) in the eastern Tethys (India and Pakistan); middle Eocene, central Tethys (Egypt); late Eocene, western Tethys (southeastern United States, Australasia); late Eocene, Morocco; and extending through the Pleistocene.

AMBULOCETIDAE† Thewissen et al. 1996

Ambulocetus† Thewissen et al. 1994

Gandakasia† Dehm and Oettingen-Spielberg 1958

Himalayacetus† Bajpal and Gingerich 1998

PAKICETIDAE† Thewissen et al. 1996

   Ichthyolestes† Dehm and Oettingen-Spielberg 1958

   Nalacetus† Thewissen and Hussain 1998

   Pakicetus† Gingerich and Russell 1981

  PROTOCETIDAE† Stromer 1908

   Aegyptocetus† Bianucci and Gingerich 2011

   Artiocetus† Gingerich et al. 2001

   Babiacetus† Trivedy and Satsangi 1984

   Carolinacetus† Geisler et al. 2005

   Crenatocetus† McLeod and Barnes 2008

   Dhedacetus† Bajpai and Thewissen 2014

   Eocetus† Fraas 1904

   Gaviacetus† Gingerich et al. 1995

   Georgiacetus† Hulbert et al. 1998

   Indocetus† Sahni and Mishra 1975

   Kharodacetus† Bajpai and Thewissen 2014

   Maiacetus† Gingerich et al. 2009

   Makaracetus† Gingerich et al. 2005

   Natchitochia† Uhen 1998

   Pappocetus† Andrews 1920

   Pontobasileus† Leidy 1873

   Protocetus† Fraas 1904

   Qaisracetus† Gingerich et al. 2001

   Rodhocetus† Gingerich et al. 1994

   Takracetus† Gingerich et al. 1995

   Togocetus† Gingerich and Cappetta 2014

  REMINGTONOCETIDAE† Kumar and Sahni 1986

   Andrewsiphius† Sahni and Mishra 1975

   Attockicetus† Thewissen and Hussain 2000

   Dalanistes† Gingerich et al. 1995

   Kutchicetus† Bajpai and Thewissen 2000

   Rayanistes† Bebej et al.2016

   Remingtonocetus† Kumar and Sahni 1986

  PELAGICETI Uhen 2008

  BASILOSAURIDAE† Cope 1867

   Ancalecetus† Gingerich and Uhen 1996

   Basilosaurus† Harlan 1834

   Basiloterus† Gingerich et al. 1997

   Basilotritus† Gol'din and Zvonok 2013

   Chrysocetus† Uhen and Gingerich 2001

   Cynthiacetus† Uhen 2005

   Dorudon† Gibbes 1845

   Masracetus† Gingerich 2007

   Ocucajea† Uhen et al. 2011

   Platyosphys† Kellogg 1936

   Pontogeneus† Leidy 1852

   Saghacetus† Gingerich 1992

   Stromerius† Gingerich 2007

   Supayacetus† Uhen et al. 2011

   Zygorhiza† True 1908

KEKENODONTIDAE† Mitchell 1989

Kekenodon† Hector 1881

NEOCETI Fordyce and Muizon 2001

ODONTOCETI Flower 1867

Content—Twenty-nine genera and 76 extant species, of which one is extinct, are currently recognized.

Fossil history—The earliest odontocetes are known from the late Oligocene (29.3–23 Ma)-Pleistocene of the western North Atlantic (southeastern United States), Parathethys (Europe, Asia), eastern North Pacific (Japan), and western South Pacific (New Zealand); early early Miocene-Pleistocene, western South Atlantic (South America) and eastern North Pacific (Oregon and Washington).

STEM ODONTOCETES† Gray 1863

Archaeodelphis† Allen 1921

Argyrocetus† Lydekker 1894

Cotylocara† Geisler et al. 2014

Huaridelphis† Lambert et al. 2014

Papahu† Aguirre-Fernandez and Fordyce 2014

AGOROPHIIDAE† Uhen 2008

Agorophius† Cope 1895

ASHLEYCETIDAE† Geisler and Sanders 2015

Ashleycetus† Geisler and Sanders 2015

EOPLATANISTIDAE† Muizon 1988

Eoplatanista† Dal Piaz 1916

MIROCETIDAE† Geisler and Sanders 2015

Mirocetus† Geisler and Sanders, 2015

PROSQUALODONTIDAE† Fordyce and Muizon 2001

Prosqualodon† Stromer 1908

SIMOCETIDAE† Fordyce 2002

Simocetus† Fordyce 2002

SQUALODONTIDAE† Brandt 1873

Squalodon† Grateloup 1840

WAIPATIIDAE† Fordyce 1994

Awamokoa† Tanaka and Fordyce 2016

Otekaikea† Tanaka and Fordyce 2014

Waipatia† Fordyce 1994

XENOROPHIDAE† Uhen 2008

Albertocetus† Uhen 2008

Echovenator† Churchill et al. 2016

Xenorophus† Kellogg 1923

PAN-PHYSETEROIDEA Vélez-Juarbe et al. 2015

Content—New clade name is to include the following fossil genera:

Acrophyseter† Lambert et al. 2008

Albicetus† Boersma and Pyenson 2015

Brygmophyseter† Kimura et al. 2006

Diaphorocetus† Trouessart 1898

Eudelphis† Du Bus 1872

Idiophyseter† Kellogg 1925

Idiorophus† Kellogg 1925

Livyatan† Lambert et al. 2010

Orycterocetus† Leidy 1853

Placoziphius† Van Beneden 1869

Zygophyseter† Bianucci and Landini 2006

PHYSETEROIDEA Gray 1868

PHYSETERIDAE Gray 1821

Content—One genus and species is recognized, *Physeter macrocephalus* Linnaeus 1758 (syn. *Physeter catadon*): sperm whale, cachalot.

Distribution—Physeterids are widely distributed in all oceans of the world, avoiding only the polar pack ice in both hemispheres.

Fossil history—Extant genus *Physeter* is recorded from the Pliocene. The record of physeterids goes back at least to the early early Miocene (23.3-21.5 Ma), Mediterranean (Italy), Paratethys (Europe), and western South Atlantic (Argentina); late early Miocene, western South Pacific (Australasia), eastern North Pacific (central California), and western North Pacific (Japan); early middle Miocene, western North Atlantic (Maryland, Virginia); late middle Miocene-late Miocene, eastern North Atlantic (western Europe); late middle Miocene-late Pliocene, eastern North Atlantic (Florida); late late Miocene, eastern South Pacific (Peru); late Pliocene, eastern North Pacific (California), and perhaps earlier if *Ferecetotherium*† is included from the late Oligocene (29-23 Ma) (see Fordyce, 2009). Extinct genera are reviewed by Bianucci and Landini (2006), Lambert et al. (2008, 2010), Boersma and Pyenson (2015), and Vélez-Juarbe et al. (2015).

Aulophyseter† Kellogg 1927

Ferecetotherium† Mchedlidze 1970

Physeterula† Van Beneden 1877

KOGIIDAE Gill 1871; Miller 1923

Content—One genus and two extant species are recognized, *Kogia breviceps*, pygmy sperm whale, and *Kogia sima*, dwarf sperm whale.

Distribution—Kogiids are distributed in all oceans of the world (temperate, subtropical, and tropical waters), with the range of *Kogia breviceps* recorded as in warmer seas (see Handley, 1966).

Fossil history—The extant genus *Kogia* is known from the late Pliocene of Italy. The fossil record of kogiids goes back to the early Miocene (20.4 Ma) eastern North Atlantic (Belgium), eastern South Pacific (Peru); early Pliocene, western North Atlantic (Florida); late Miocene, Caribbean (Panama); late Pliocene, Mediterranean (Italy). Extinct genera are reviewed by Vélez-Juarbe et al. (2015).

Aprixokogia† Whitmore and Kaltenbach 2008

Nanokogia† Vélez-Juarbe et al. 2015

Praekogia† Barnes 1973

Scaphokogia† Muizon 1988

Thalassocetus† Abel 1905

SYNRHINA Geisler et al. 2011

ZIPHIIDAE Gray 1865

Content—Six genera and 22 extant species are known.

Distribution—Ziphiids occur in temperate and tropical waters of all oceans, with some species located far offshore in deep waters (exceptions are *Berardius bairdii* Stejneger 1883 from the North Pacific and *Mesoplodon bidens* Sowerby 1804 from cold-temperature to subarctic North Atlantic).

Fossil history—The extant genus *Mesoplodon* is known from the late middle Miocene (14.6-11

Ma), eastern North Atlantic (western Europe); late Pleistocene-Holocene, western South Pacific (Argentina). The record of ziphiids extends back to the early Miocene (23.3-21.5 Ma), eastern North Pacific (Oregon, Washington); late middle Miocene (14.6-11 Ma), eastern North Atlantic (western Europe); early Pliocene, North America; middle-late Miocene, western North Pacific (Japan); middle to late Miocene-early Pliocene, Australasia, western North Atlantic (Virginia, North Carolina, Florida), eastern North Atlantic (Denmark), eastern South Pacific (Peru), and southwest Atlantic (Argentina); late Pliocene, Mediterranean (Italy) and Australasia. Extinct taxa are reviewed by Fordyce (2009), Bianucci et al. (2013b, 2016), and Lambert et al. (2013).

Africanacetus† Gol'din and Vishnyakova 2013

Anoplonassa† Cope 1869

Aporotus† Du Bus 1868

Archaeoziphius† Lambert and Louwye 2006

Belemnoziphius† Huxley 1864

Beneziphius† Lambert 2005

Caviziphius† Bianucci and Post 2005

Chavinziphius† Bianucci et al. 2016

Chimuziphius† Bianucci et al. 2016

Choneziphius† Duvernoy 1851

Dagonodum† Ramassamy 2016

Eboroziphius† Leidy 1876

Globicetus† Bianucci et al. 2013

Imocetus† Bianucci et al. 2013

Ihlengesi† Bianucci, Lambert, and Post 2007

Izikoziphius† Buono and Cozzuol 2013

Messapicetus† Bianucci et al. 1992

Microberardius† Bianucci, Lambert, and Post 2007

Nazcacetus† Lambert et al. 2009

Nenga† Bianucci et al. 2007

Ninoziphius† Muizon 1983

Notoziphius† Buono and Cozzuol 2013

Pterocetus† Bianucci et al. 2007

Tusciziphius† Bianucci 1997

Xhosacetus† Bianucci et al. 2007

Ziphirostrum† Du Bus 1868

PLATANISTOIDEA fide Boersma and Pyenson 2016

Content—The phylogenetic position of a number of stem odontocetes, many allied in the grade-level taxon Platanistoidea, has been controversial, and some systematic analyses do not find support for monophyly of this group (see Geisler et al., 2011; also see the text discussion). This group has included, among others, *Dalpiazina*† Muizon 1988, *Ninjadelphis*† Kimura and Barnes 2016, *Notocetus*† Moreno 1892, *Zarhinocetus*† Barnes and Reynolds 2009, *Huaridelphis*† Lambert et al. 2014, *Goedertius*† Kimura and Barnes 2016, *Potamodelphis*† Allen 1921, and *Zarhachis*† Cope 1868. Recent comprehensive phylogenetic analysis (see Boersma and Pyenson, 2016) proposes that this clade includes Platanistidae and fossil taxa Waipatiidae, Allodelphinidae, and Squalodelphinidae.

ALLODELPHINIDAE Boersma and Pyenson 2016

Allodelphis† Wilson 1935

Arktocara† Boersma and Pyenson 2016

    Goedertius† Kimura and Barnes 2016

    Ninjadelphis† Kimura and Barnes 2016

    Zarhinocetus† Barnes and Reynolds 2009

   PLATANISTIDAE Gray 1863

Content—A single genus and extant species, *Platanista gangetica* Zebeck 1801, and two subspecies (*P. g. gangetica* Zebeck 1801 and *P. g. minor* Owen 1853) are recognized, following Perrin and Brownell (2001).

Distribution—Platanistids occur in the Indus and Ganges River drainages of Pakistan and India.

Fossil history—Fossil platanistids are known from the late early Miocene (21.5-16.3 Ma), western North Atlantic (Florida), and eastern North Atlantic (Europe—Germany, South America—Venezuela); early middle Miocene, western North Atlantic (Maryland, Virginia, Florida). The diversification of platanistids is further discussed by Geisler et al. (2011), Bianucci et al. (2013a), and Boersma and Pyenson (2016). Several fossil taxa, including *Pachyacanthus*† Brandt 1871 and *Prepomatodelphis*† Barnes 2002, formerly recognized in this family are not included in the latest concept.

    Pomatodelphis† Allen 1921

    Zarhachis† Cope 1868

   SQUALODELPHINIDAE† Ginsberg and Janvier 1971

    Notocetus†

  DELPHINIDA Muizon 1988

   KENTRIODONTIDAE† Slijper 1936

    Belonodelphis† Muizon 1988

    Delphinodon† Leidy 1869

    Hadrodelphis† Kellogg 1966

    Heterodelphis† Brandt 1873

    Incacetus† Colbert 1944

    Kampholophos† Rensberger 1969

    Kentriodon† Kellogg 1927

    Leptodelphis† Kirpichnikov 1954

    Liolithax† Kellogg 1931

    Lophocetus† Cope 1867

    Macrokentriodon† Dawson 1996

    Microphocaena† Kudrin and Tatarinov 1965

    Pithanodelphis† Abel 1905

    Rudicetus† Bianucci 2001

    Sarmatodelphis† Kirpichnikov 1954

    Sophianacetus† Kazar 2006

    Tagicetus† Lambert et al. 2005

   LIPOTIDAE Zhou Qian and Li 1979

Content—A single recently extinct genus and extant species is recognized, *Lipotes vexillifer* Miller 1918 baji, Yangtze river dolphin.

Distribution—Lipotes is confined to Yangtze (Changjiang) River drainage on the Chinese mainland.

Fossil history—The extinct genus *Prolipotes*, reported from the Neogene of China (see Zhou et al., 1984), has not been confirmed as a lipotiid due to its incompleteness (see Fordyce, 2009). *Parapontoporia*†, originally referred to the Pontoporiidae, has since been placed as sister group to *Lipotes* (e.g., Geisler et al., 2011, 2012).

    Parapontoporia† Barnes 1984

INIOIDEA Muizon 1988

FAMILY *INCERTAE SEDIS*

Brujadelphis† Lambert et al. 2017

INIIDAE Flower 1867

Content—A single genus, *Inia* d'Orbigny 1834, and three extant species and two subspecies are recognized.

Distribution—Iniids occur in the Amazon and Orinoco River drainages in central and northern South America.

Fossil history—Iniids are known from the early late Miocene (10.4-6.7 Ma), western South Atlantic (Argentina, Uruguay); early Pliocene, western North Atlantic (Florida). Fossil iniids are reviewed by Fordyce (2009), Geisler et al. (2012), and Pyenson et al. (2015).

Goniodelphis† Allen 1941

Ischyrorhynchus† Ameghino 1891

Isthminia† Pyenson et al. 2015

Meherrinia† Geisler et al. 2012

Saurocetes† Burmeister 1871

PONTOPORIIDAE Gill 1871; Kasuya 1973

Content—A single extant genus and species is recognized, *Pontoporia blainvillei* Franciscana, La Plata dolphin.

Distribution—Pontoporiids are restricted to coastal central Atlantic waters of South America.

Fossil history—The fossil record of pontoporiids extends back to the late Miocene (10.4-6.7 Ma), western South Atlantic (Argentina); late late Miocene, eastern North Pacific (California, Baja California); Miocene, North Sea and eastern South Pacific (Peru); early Pliocene, western North Atlantic (Virginia, North Carolina), eastern North Pacific (California), and eastern South Pacific (Peru, Chile). *Pontoporia* is reported from the early late Miocene (10.4-6.4 Ma), western South Atlantic (Argentina). Extinct genera are reviewed most recently by Geisler et al. (2012) and Lambert and Muizon (2013). *Auroracetus* and *Stenasodelphis* are identified as inioids by Pyenson et al. (2015).

Auroracetus† Gibson and Geisler 2009

Brachydelphis† Muizon 1988

Pliopontos† Muizon 1983

Pontistes† Burmeister 1885

Protophocoena† Abel 1905

Stenasodelphis† Godfrey and Barnes 2008

DELPHINOIDEA Flower 1865

ALBIREONIDAE† Barnes 1984

Albireo† Barnes 1984

DELPHINIDAE Gray 1821

Content—Seventeen genera and 38 extant species are recognized.

Distribution—Delphinids are widely distributed in all oceans of the world.

Fossil history—Extant genera *Globicephala* and *Pseudorca* are known from the late Pleistocene (10,000-790 years ago), eastern North Atlantic (Florida) (Morgan, 1994). *Delphinus* and *Tursiops* are known from the late Pleistocene, eastern North Atlantic (western Europe) (see Van der Feen, 1968). *Tursiops* is also reported from the Pleistocene of Japan (see Kimura et al., 2012). *Delphinus* is known from the late Pliocene, New Zealand (see Fordyce, 1991; McKee, 1994).

The oldest known delphinids are from the latest Miocene (11-7 Ma)-Pleistocene, western North Atlantic (Virginia, North Carolina); late Miocene-Pleistocene, western North Pacific

(New Zealand, Japan); late Miocene-early Pliocene, eastern North Pacific (California); early Pliocene (western Europe); early Pliocene, eastern South Pacific (Peru); late Pleistocene, eastern North Atlantic (western Europe) (see Bianucci et al., 2009; Murakami et al., 2014). Extinct genera are reviewed by Fordyce (2009) and Bianucci et al. (2009, 2013b).

Arimidelphis† Bianucci 2005

Astadelphis† Bianucci 1996

Australodelphis† Fordyce et al. 2002

Eodelphinus† Murakami et al. 2014

Etruridelphis† Bianucci et al. 2009

Hemisyntrachelus† Brandt 1873

Lagenorhynchus† Gray 1846

Platalearostrum† Post and Kompanje 2010

Protoglobicephala† Aguirre-Fernandez et al. 2009

Septidelphis† Bianucci 2013

MONODONTOIDAE Geisler et al. 2011

MONODONTIDAE Gray 1821; Miller and Kellogg 1955

Content—Two extant genera, *Monodon* Linnaeus 1758 and *Delphinapterus* Pallas 1776, and two species are recognized.

Distribution—The beluga and narwhal are confined to Arctic and sub-Arctic waters.

Fossil history—The extant genus *Delphinapterus* is known from the Pliocene (5.3 Ma), western North Atlantic (Virginia, North Carolina) (see Whitmore, 1994). The extant species *Delphinapterus leucas* is known from the late Pleistocene (781 kya), western North Atlantic (Canada) (see Harrington, 1977). Monodontids are known in the fossil record as far back as early late Miocene (10.4-6.7 Ma); late late Miocene, eastern North Pacific (California); late late Miocene, eastern North Pacific (Baja California); latest Miocene-early Pliocene, western North Atlantic (Virginia, North Carolina) and eastern North Pacific (California); late Pliocene, eastern North Pacific (California); late Pleistocene, eastern North Atlantic (western Europe) and Arctic Ocean (northern Alaska). Extinct genera are reviewed by Fordyce (2009) and Vélez-Juarbe and Pyenson (2012).

Bohaskaia† Vélez-Juarbe and Pyenson 2012

Denebola† Barnes 1984

PHOCOENIDAE Gray 1825; Bravard 1885

Content—Three extant genera, 7 species, and 8 subspecies are recognized (Rosel et al., 1995; Wang et al., 2008; Jefferson and Wang, 2011).

Distribution—Phocoenids are distributed throughout the world.

Fossil history—Phocoenids are known in the fossil record from early late Miocene-late late Miocene (10.4-6.7 Ma), eastern North Pacific (California); late late Miocene, eastern North Pacific (Baja California) and eastern South Pacific (Peru); early Pliocene, western North Atlantic (Virginia, North Carolina); early-late Pliocene, eastern North Pacific (California); late Pleistocene, eastern North Atlantic (western Europe) and western North Atlantic (Canada). Extinct genera are reviewed by Murakami et al. (2012a, b; 2014). The extant species *Phocoena phocoena* Linnaeus 1758 is known from the late Pleistocene, eastern North Atlantic (western Europe) (see Van der Feen, 1968) and western North Atlantic (Canada) (see Harrington, 1977).

Archaeophocoena† Murakami et al. 2012

Australithax† Muizon 1988

Brabocetus† Colpaert et al. 2015

Haborophocoena† Ichishima and Kimura 2005

Lomacetus† Muizon 1986

Loxolithax† Kellogg 1931

Miophocoena† Murakami et al. 2012

Numataphocoena† Ichishima and Kimura 2000

Piscolithax† Muizon 1983

Pterophocoena† Murakami et al. 2012

Salumiphocaena† Barnes 1985

Semirostrum† Racicot et al. 2014

Septemtriocetus† Lambert 2008

ODOBENOCETOPSIDAE† Muizon 1993

Fossil history—Odobenocetopsids are known from the Miocene (7.246-5.332 Ma) of Chile and Peru.

Odobenocetops† Muizon 1993

EURHINODELPHINIDAE† Abel 1901

Ceterhinops† Leidy 1877

Eurhinodelphis† Van Beneden and Gervais 1880

Iniopsis† Lydekker 1893

Mycteriacetus† Lambert 2004

Phocaenopsis† Huxley 1859

Schizodelphis† Gervais 1861

Vanbreenia† Bianucci and Landini 2002

Xiphiacetus† Lambert 2005

Ziphiodelphis† Dal Piaz 1908

FAMILY *INCERTAE SEDIS* (fide Lambert et al. 2015)

Argyrocetus† Lydekker 1894

Chilcacetus† Lambert et al. 2015

Macrodelphinus† Wilson 193

Squaloziphius† Muizon 1991

MYSTICETI Flower 1864

Content—Six genera and 14 extant species are currently recognized.

Fossil history—The earliest mysticetes are known from the late Eocene-early Oligocene (34-33 Ma), Antarctica; late Oligocene-Pleistocene, eastern North Pacific (California) and western North Pacific (Japan, Australasia); early early Miocene-Pleistocene, western South Atlantic; early middle Miocene, western North Atlantic; late middle Miocene-Pleistocene, eastern North Atlantic (western Europe) (Deméré et al., 2005).

LLANOCETIDAE† Mitchell 1989

Llanocetus† Mitchell 1989

MAMMALODONTIDAE† Mitchell 1989

Janjucetus† Fitzgerald 2006

Mammalodon† Pritchard 1939

AETIOCETIDAE† Emlong 1966

Aetiocetus† Emlong 1966

Ashorocetus† Barnes et al. 1995

Chonecetus† Russell 1968

Fucaia† Marx et al. 2015

Morawanocetus† Barnes et al. 1995

Willungacetus† Pledge 2005

CHAEOMYSTICETI† Mitchell 1989

Halicetus† Kellogg 1969

Piscocetus† Pilleri and Pilleri 1989

Sitsqwayk† Peredo and Uhen 2016

EOMYSTICETIDAE† Sanders and Barnes 2002

Eomysticetus† Sanders and Barnes 2002

Matapanui† Boessenecker and Fordyce 2016

Micromysticetus† Sanders and Barnes 2002

Tohoraata† Boessenecker and Fordyce 2014

Tokarahia† Boessenecker and Fordyce 2015

Waharoa† Boessenecker and Fordyce 2015

Whakakai† Tsai and Fordyce 2016

Yamatocetus† Okazaki 2012

BALAENOMORPHA Geisler and Sanders 2003

BALAENIDAE Gray 1825

Content—Two genera and three extant species are recognized.

Distribution—Bowheads are circumpolar in the Arctic, and right whales are known in temperate waters of both hemispheres.

Fossil history—Balaenids are known in the fossil record from the early late Miocene (23 Ma), western North Atlantic (Florida); late late Miocene, eastern North Pacific (California); late Miocene-early Pliocene, western South Pacific (Australasia); early Pliocene, western North Atlantic (Virginia, North Carolina), eastern North Pacific (California), and western North Pacific (Japan); late Pliocene, eastern North Pacific (California), eastern North Pacific (Florida), and Mediterranean (Italy) (see Churchill et al., 2012).The genus *Balaena* Linnaeus 1758 is reported from the early Pliocene (5.2-3.4 Ma), western North Atlantic (Virginia, North Carolina); late Pliocene, Mediterranean (Italy) and western North Pacific (Japan); early Pleistocene, eastern North Pacific (Oregon). Extant species *Balaena mysticetus* is reported from the late Pleistocene, North Atlantic (Canada) and North Atlantic (Sweden). The genus *Eubalaena* Gray 1864 is reported from the late Miocene-early Pliocene, western North Pacific (Japan). *Eubalaena glacialis* is known from the late Pleistocene, western North Atlantic (Florida).

Remarks—The taxonomic history of fossil balaenids has been reviewed by Churchill et al. (2012).

Balaenella† Bisconti 2005

Balaenula† Van Beneden 1872

Idiocetus† Capellini 1876

Morenocetus† Cabrera 1926

Peripolocetus† Kellogg 1931

PLICOGULAE Geisler et al. 2011

NEOBALAENIDAE Gray 1874; Miller 1923

Content—One extant genus and species is recognized, *Caperea marginata*, pygmy right whale.

Distribution—*C. marginata* is known only in temperate waters of the southern hemisphere (including Tasmania and the coasts of South Australia, New Zealand, and South Africa).

Fossil history—In addition to a possible neobalaenid from the Miocene of Australia (see Fitzgerald, 2012), the first well-documented record of a fossil neobalenid, *Miocaperea pulchra*, has been described from the late Miocene (11.608-7.246 Ma) of Peru (see Bisconti, 2012).

Remarks—According to Fordyce and Marx (2013), the pygmy right whale may be the last survivor of the extinct family Cetotheriidae.

Miocaperea† Bisconti 2012

THALASSOTHERII Bisconti et al. 2013

Cetotheriops† Brandt 1871

Isocetus† Van Beneden 1880

CETOTHERIIDAE† *sensu stricto*

Brandtocetus† Gol'din and Startsev 2014

Cephalotropsis† Cope 1896

Cetotherium† Brandt 1843

Eucetotherium† Brandt 1873 (syn. *Vampalus*†)

Herentalia† Bisconti 2015

Herpetocetus† Van Beneden 1872

Hibacetus† Otsuka and Ota 2008

Imerocetus† Mchedildze 1964

Joumocetus† Kimura and Hasegawa 2010

Kurdalogonus† Taarasenko and Lopatin 2012

Metopocetus† Cope 1896

Nannocetus† Kellogg 1929

Pinocetus† Czyzewska and Ryziewicz 1976

Piscobalaena† Pilleri and Siber 1989

Zygiocetus† Tarasenoko 2014

CETOTHERIIDAE† *sensu lato* Boessenecker and Fordyce 2015

Aglaocetus† Kellogg 1934

Cephalotropsis† Cope 1896

Cophocetus† Packard and Kellogg 1934

Diorocetus† Kellogg 1968

Isanacetus† Kimura and Ozawa 2002

Mauicetus† Benham 1939

Otradnocetus† Mchedlidze 1984

Parietobalaena† Kellogg 1924

Pelocetus† Kellogg 1965

Thinocetus† Kellogg 1969

Tiphyocetus† Kellogg 1931

Titanocetus† Bisconti 2006

Uranocetus† Steeman 2009

TRANATOCETIDAE† Gol'din and Steeman 2015

"*Aulocetus*"† Van Beneden 1875

"*Cetotherium*"† Brandt 1843

Mesocetus† Van Beneden 1880

Mixocetus† Kellogg 1934

Plesicetopsis† Brandt 1873

Tranatocetus† Gol'din and Steeman 2015

BALAENOPTEROIDEA Gray 1868

ESCHRICHTIIDAE Ellerman and Morrison-Scott 1951

Content—One genus and extant species (*Eschrichtius robustus*) is known.

Distribution—Three or more stocks of gray whales are recognized: North Atlantic stock (or stocks), apparently exterminated in recent historical times (perhaps as late as the seventeenth or eighteenth century); a Korean or western Pacific stock hunted at least until 1966 and now close to extinction; and a California or eastern Pacific stock (see Jefferson et al., 2015).

Fossil history—The genus *Eschrichtius* Gray 1864 is known from the late Pliocene, western North Pacific (Japan); late Pleistocene, western North Atlantic (Florida, Georgia) and eastern North Pacific (California); Quaternary (11-10 kya), Taiwan. Extinct genera are reviewed by Bisconti (2008).

Archaeschrichtius† Bisconti and Varola 2006

Diunatans† Bosselaers and Post 2010

Eschrichtioides† Bisconti 2008

Gricetoides† Whitmore and Kaltenbach 2008

Plesiobalaenoptera† Bisconti 2010

BALAENOPTERIDAE Gray 1864

Content—Two extant genera (*Balaenoptera*, *Megaptera*) containing eight species are recognized.

Distribution—Humpback, minke, fin, and blue whales are widely distributed in all oceans. Sei whales and Bryde's whales are usually found in tropical and warm temperate waters, not in polar regions as are other rorquals.

Fossil history—Balaenopterids are known from the early late Miocene (12 Ma), eastern North Pacific (California); late late Miocene, western North Atlantic (Maryland, Virginia), eastern North Pacific (southern California, Baja California), western North Pacific (Japan), and eastern South Pacific (Peru); latest Miocene-early Pliocene, Australasia; early Pliocene, eastern North Atlantic (western Europe), western North Atlantic (Virginia, North Carolina), western North Atlantic (Florida), eastern North Pacific (California), western North Pacific (Japan), and eastern South Pacific (Peru); late Pliocene, Mediterranean (Italy), western North Atlantic (Florida), eastern North Pacific (California), and western South Pacific (Australasia); early Pleistocene, western North Atlantic (western Europe); late Pleistocene, western North Atlantic (Canada) and eastern North Pacific (California).

The genus *Megaptera* is known from the early late Miocene, eastern North Pacific (central California); late Miocene-early Pliocene (Australasia); early Pliocene, western North Atlantic (Virginia, North Carolina); late Pliocene, western North Atlantic (Florida); late Pleistocene, western North Atlantic (Florida), and western North Atlantic (Canada). Sei whale (*Balaenoptera borealis*) is reported from the late Pleistocene, western North Pacific (Japan). The humpback whale (*Megaptera novaeangliae*) is known from the late Pleistocene, western North Atlantic (Florida), and western North Atlantic (Canada). Extinct genera are reviewed by Demére et al. (2005) and Marx and Kohno (2016).

Archaebalaenoptera† Bisconti 2007

Cetotheriophanes† Brandt 1873

Eobalaenoptera† Dooley et al. 2004

Fragilicetus† Bisconti and Bosselaers 2016

Incakujira† Marx and Kohno 2016

Notiocetus† Ameghino 1891

Palaeocetus† Seeley 1865

Parabalaenoptera† Zeigler et al. 1997

Plesiocetus† Van Beneden 1859

Praemegaptera† Behrmann 1995

Protororqualus† Bisconti 2007

SIRENIA Illiger 1811

Content—This group is composed of two families containing all living sirenians.

Distribution—Sirenians occur in the Indo-Pacific (dugongs) and the southeastern United States, Caribbean, and the Amazon drainage of South America (manatees).

Fossil history—Sirenians are recognized in the fossil record beginning in the early Eocene

(56.5-50 Ma), western Tethys (Jamaica); middle Eocene, eastern Tethys (Pakistan, India), central Tethys (Egypt), and western Tethys (southeastern United States); late Eocene, central Tethys (Egypt, Europe); early middle Miocene, Australasia (New Guinea); and extending through to the Holocene (for updates, see Vélez-Juarbe and Domning, 2014a, b).

  PRORASTOMIDAE† Cope 1889

   Pezosiren† Domning 2001

   Prorastomus† Owen 1855

  PROTOSIRENIDAE† Sickenberg 1934

   Ashokia† Bajpai et al. 2009

   Protosiren† Abel 1904

  DUGONGIDAE Gray 1821

Content—Two extant genera (*Dugong, Trichechus*) and two species are recognized. The taxonomic history of dugongids is reviewed by Domning (1996), but see Vélez-Juarbe et al. (2012) and Vélez-Juarbe and Domning (2014a, b) for updates.

Distribution—The dugong is widely distributed in shallow coastal bays of the Indo-Pacific region.

Fossil history—The extant species *Dugong dugon* Illiger 1811 may also occur in the Pleistocene (see Domning, 1996). Fossil dugongids are reported from the middle Eocene (50-38.6 Ma), central Tethys; late Eocene, central Tethys (Egypt); late Eocene, Morocco; early Oligocene, western Atlantic and Caribbean (Puerto Rico); late Oligocene, Parathethys (Caucasia) and central Tethys (Europe); early early Miocene, western South Atlantic (Argentina); late early Miocene, Parathethys (Austria, Switzerland), western North Atlantic (Florida), and western South Atlantic (Brazil); early middle Miocene, western North Atlantic (Maryland, Virginia); late middle Miocene, Mediterranean (Italy), eastern North Atlantic (western Europe), western North Atlantic (Florida), and eastern North Pacific (California, Baja California); early late Miocene, western North Atlantic (Florida), eastern North Pacific (California), and western North Pacific (Japan); late late Miocene, eastern North Pacific (southern California) and western North Pacific (Japan); early Pliocene, western North Atlantic (Florida), eastern North Pacific (California), and western North Pacific (Japan); late Pliocene, Mediterranean (Italy), western North Atlantic (Florida), and eastern North Pacific (California); early Pleistocene, western South Pacific (Australasia); late Pleistocene, eastern North Pacific (Alaska to central California) and western North Pacific (Japan).

   Anisosiren† Kordos 1979

   Eosiren† Andrews 1902

   Eotheroides† Palmer 1899

   Indosiren† von Koenigswald 1952

   Metaxytherium† de Christol 1840

   Miodugong† Deraniyagala 1969

   Paralitherium† Kordos 1977

   Priscosiren† Vélez-Juarbe and Domning 2014

   Prohalicore† Flot 1887

   Prototherium† de Zigno 1887

   Sirenavus† Kretzoi 1941

     Dugonginae Simpson 1932

       Bharatisiren† Bajpai and Domning 1997

       Callistosiren† Vélez-Juarbe and Domning 2015

       Corystosiren† Domning 1990

> > *Crenatosiren*† Domning 1991
> >
> > *Dioplotherium*† Cope 1883
> >
> > *Domningia*† Thewissen and Bajpai 2009
> >
> > *"Halitherium"*† Kaup 1838
> >
> > *Kutchisiren*† Bajpai et al. 2010
> >
> > *Nanosiren*† Domning and Aguilera 2008
> >
> > *Rytiodus*† Lartet 1866
> >
> > *Xenosiren*† Domning 1989
> >
> > **Hydrodamalinae** Palmer 1895
> >
> > > *Dusisiren*† Domning 1978
> > >
> > > *Hydrodamalis*† Retzius 1794

TRICHECHIDAE Gill 1872

Content—One extant genus (*Trichechus* Linnaeus 1758) and three species are recognized. Domning and Hayek (1986) described the morphological distinction of subspecies; see also Domning (2005) for discussion of morphologically distinct late Pleistocene subspecies *Trichechus manatus bakerorum*†.

Distribution—Members of this family inhabit the Gulf coast of Florida (*Trichechus manatus manatus*), the Caribbean and South American coast (*T. manatus latirostris*), the Amazon Basin (*Trichechus inunguis*), and West Africa (*Trichechus senegalensis*).

Fossil history—The Antillean manatee *Trichechus manatus* is reported from the late Pleistocene, western North Atlantic (Florida). Fossil trichechids are reported from the late Miocene, western South Pacific (Argentina) and eastern South Pacific (Brazil, Peru); early Pliocene, western North Atlantic (Virginia, North Carolina); early-late Pleistocene, western North Atlantic (Florida) (see Cozzuol, 1996; South American record).

> *Anomotherium*† Siegfried 1965
>
> *Miosiren*† Dollo 1889
>
> *Potamosiren*† Reinhart 1951
>
> *Ribodon*† Ameghino 1883

DESMOSTYLIA† Reinhart 1953

Content—There are no living genera.

Fossil history—Desmostylians are known from the late Oligocene-middle Miocene, North America and Japan (see Domning, 1996; Beatty, 2009; Barnes, 2013; Beatty and Cockburn, 2015).

Ashoroa† Inuzuka 2000

Behemotops† Domning et al. 1986

Cornwallius† Hay 1923

Desmostylus† Nagao 1937

Kronokotherium† Pronina 1957

Ounalashkastylus† Chiba et al. 2015

Seuku† Beatty and Cockburn 2015

Vanderhoofius† Reinhart 1959

PALEOPARADOXIIDAE† Reinhart 1959

> *Archaeoparadoxia*† Barnes 2013
>
> *Neoparadoxia*† Barnes 2013
>
> *Paleoparadoxia*† Reinhart 1959

References

Amson, E., Muizon, C. de. 2013. A new durophagous phocid (Mammalia: Carnivora) from the late Neogene of Peru and considerations on monachine seals phylogeny. *Journal of Systematic Paleontology* 12: 523-548.

Barnes, L.G. 2013. A new genus and species of late Miocene paleoparadoxiid (Mammalia, Desmostylia) from California. *Natural History Museum of Los Angeles County Contributions in Science* 521: 51-114.

Beatty, B.L. 2009. New material of *Cornwallius sookensis* (Mammalia: Desmostylia) from the Yaquina Formation of Oregon. *Journal of Vertebrate Paleontology* 29: 894-909.

Beatty, B.L., Cockburn, T.C. 2015. New insights on the most primitive desmostylians from a partial skeleton of *Behemotops* (Desmostylia, Mammalia) from Vancouver Island, British Columbia. *Journal of Vertebrate Paleontology*, doi: 10.1080/02724634.2015.979939.

Berta, A. 2009. Pinniped evolution. In: Perrin, W.F., Wursig, B., Thewissen, J.G.M. (eds.), *Encyclopedia of Marine Mammals*, 2nd ed. Elsevier, San Diego, CA, pp. 861-868.

Berta, A., Churchill, M. 2012. Pinniped taxonomy: review of currently recognized species and subspecies, and evidence used for their description. *Mammal Reviews* 42: 207-234.

Berta, A., Ray, C.E., Wyss, A.R. 1989. Skeleton of the oldest known pinniped, *Enaliarctos mealsi. Science* 244: 60-62.

Bianucci, G. 2013. *Septidelphis morii*, n. gen. et sp. from the Pliocene of Italy: new evidence of the explosive radiation of true dolphins (Odontoceti, Delphinidae). *Journal of Vertebrate Paleontology* 33: 722-740.

Bianucci, G., Landini, W. 2006. Killer sperm whale: a new basal physeteroid (Mammalia, Cetacea) from the Late Miocene of Italy. *Zoological Journal of the Linnean Society* 148: 103-131.

Bianucci, G., Vaiani, S.C., Casati, S. 2009. A new delphinid record (Odontoceti, Cetacea) from the Early Pliocene of Tuscany (Central Italy): systematics and biostratigraphic considerations. *Neues Jahrbuch Geologie und Paläontologie* 254: 275-292.

Bianucci, G., Lambert, O., Salas-Gismondi, R., Tejada, J., Pujos, F., Urbina, M., Antoine, P.-O. 2013a. A Miocene relative of the Ganges river dolphin (Odontoceti, Platanistidae) from the Amazonian Basin. *Journal of Vertebrate Paleontology* 33(3): 741-745.

Bianucci, G., Mijan, I., Lambert, O., Post, K., Mateus, O. 2013b. Bizarre fossil beaked whales (Odontoceti, Ziphi-idae) fished from the Atlantic Ocean floor off the Iberian Peninsula. *Geodiversitas* 35: 105-153.

Bianucci, G., di Celma, C., Urbina, M., Lambert, O. 2016. New beaked whales from the late Miocene of Peru and evidence for convergent evolution in stem and crown Ziphiidae (Cetacea, Odontoceti). *PeerJ* 4: e2479. doi: 10.7717/peerj.2479.

Bisconti, M. 2008. Morphology and phylogenetic relationships of a new eschrichtiid genus (Cetacea: Mysticeti) from the Early Pliocene of northern Italy. *Zoological Journal of the Linnean Society* 153: 161-186.

———. 2012. Comparative osteology and phylogenetic relationships of *Miocaperea pulchra*, the first fossil pygmy right whale genus and species (Cetacea, Mysticeti, Neobalaenidae). *Zoological Journal of the Linnean Society* 166: 876-911.

Boersma, A.T., Pyenson, N.D. 2015. *Albicetus oxymycterus*, a new generic name and redescription of a basal physeteroid (Mammalia, Cetacea) from the Miocene of California, and the evolution of body size in sperm whales. *PLOS ONE* 10(12): e0135551.

———. 2016. *Arktocara yakataga*, a new fossil odontocete (mammalia, cetacean) from the Oligocene of Alaska and the antiquity of Platanistoidea. *PeerJ*, doi: 10.7717 /peerj.2321.

Boessenecker, R.W., Churchill, M. 2013. A reevaluation of the morphology, paleoecology, and phylogenetic relationships of the enigmatic walrus *Pelagiarctos. PLOS ONE* 8: e54311.

———. 2015. The oldest known fur seal. *Biology Letters* 11(2): 20140835.

Churchill, M., Berta, A., Deméré, T.A. 2012. The systematics of right whales (Mysticeti: Balaenidae). *Marine Mammal Science* 28: 497-521.

Churchill, M., Boessenecker, R.W., Clementz, M.T. 2014. Colonization of the Southern Hemisphere by fur seals and sea lions (Carnivora: Otariidae) revealed by combined evidence phylogenetic and Bayesian biogeographical analysis. *Zoological Journal of the Linnean Society* 172(1): 200-225.

Committee on Taxonomy, Society for Marine Mammalogy. 2016. List of Marine Mammal Species and Subspecies. www.marinemammalscience.org (consulted in July 2016).

Cozzuol, M.A. 1996. The record of the aquatic mammals in southern South America. *Müncher Geowissenschaften Abhandlungen* 30: 321-342.

Deméré, T.A., Berta, A., Adam, P.J. 2003. Pinnipedimorph

evolutionary biogeography. *Bulletin of the American Museum of Natural History* 279: 32-76.

Deméré, T.A., Berta, A., McGowen, R. 2005. The taxonomic and evolutionary history of fossil and modern balaenopteroid mysticetes. *Journal of Mammalian Evolution* 12: 99-143.

Domning, D.P. 1996. Bibliography and index of the Sirenia and Desmostylia. *Smithsonian Contributions in Paleobiology* 80: 1-611.

———. 2005. Fossil Sirenia of the West Atlantic and Caribbean region, VII: Pleistocene *Trichechus manatus* Linnaeus, 1758. *Journal of Vertebrate Paleontology* 25: 685-701.

Domning, D.P., Hayek, L.C. 1986. Interspecific and intraspecific morphological variation in manatees (Sirenia: *Trichechus*). *Marine Mammal Science* 2: 87-144.

Fitzgerald, E.M.G. 2012. Possible neobalaenid from the Miocene of Australia implies a long evolutionary history for the pygmy right whale *Caperea marginata* (Cetacea, Mysticeti). *Journal of Vertebrate Paleontology* 32: 976-980.

Fordyce, R.E. 1991. The fossil vertebrate record of New Zealand. In: Vickers-Rich, P., Monaghan, J.M., Baird, R.F., Rich, T. (eds.), *Vertebrate Paleontology of Australasia*. Pioneer Design Studio, Melbourne, Australia, pp. 1191-1314.

———. 2009. Cetacean fossil record. In: Perrin, W.F., Wursig, B., Thewissen, J.G.M. (eds.), *Encyclopedia of Marine Mammals*, 2nd. ed. Elsevier, San Diego, CA, pp. 201-207.

Fordyce, R.E., Marx, F.G. 2013. The pygmy right whale *Caperea marginata*: the last of the cetotheres. *Proceedings of the Royal Society of London B* 280(1753): 1-6.

Geisler, J.H., McGowen, M.R., Yang, G., Gatesy, J. 2011. A supermatrix analysis of genomic, morphological and paleontological data from crown Cetacea. *BMC Evolutionary Biology* 11: contrib. 112. doi: 10.1186/1471-2148-11-112.

Geisler, J.H., Godfrey, S.J., Lambert, O. 2012. A new genus and species of late Miocene inioid (Cetacea, Odontoceti) from the Meherrin River, North Carolina, USA. *Journal of Vertebrate Paleontology* 32: 198-2011.

Gradstein, F.M., Ogg, J.G., Schmitz, M.D., Ogg, G.M. 2012. *The Geologic Time Scale*. Elsevier, Oxford.

Handley, C.O. Jr. 1966. A synopsis of the genus *Kogia* (pygmy sperm whales). In: Norris, K.S. (ed.), *Whales, Dolphins, and Porpoises*. University of California Press, Berkeley, CA, pp. 62-69.

Harrington, C.R. 1977. Marine mammals from the Champlain Sea and the Great Lakes. *Annals of the New York Academy of Sciences* 288, 508-537.

IUCN. 2015. IUCN Red List of Threatened Species. Version 2015-4. www.iucnredlist.org.

Jefferson, T.A., Wang, J.Y. 2011. Revision of the taxonomy of finless porpoises (genus *Neophocaena*): the existence of two species. *JMATE* 4: 3-16.

Jefferson, T.A., Webber, M.A., Pitman, R.L. 2015. *Marine Mammals of the World*, 2nd ed. Elsevier, San Diego, CA.

Kimura, T., Yuji, T., Koichi, Y. 2012. A fossil delphinid (Cetacea, Odontoceti) from the Pleistocene Ichijiku Formation, Kazusa Group, Chiba, Japan. *Bulletin of the Gunma Museum of Natural History* 16: 71-76.

Koretsky, I.A., Domning, D.P. 2014. One of the oldest seals (Carnivora, Phocidae) from the Old World. *Journal of Vertebrate Paleontology* 34: 224-229.

Koretsky, I., Sanders, A.E. 2002. Paleontology from the late Oligocene Ashley and Chandler Bridge Formations of South Carolina 1: Paleogene pinniped remains: the oldest known seal (Carnivora: Phocidae). *Smithsonian Contributions in Paleobiology* 93: 179-183.

Lambert, O., Bianucci, G., Muizon, C. de. 2008. A new stem-sperm whale (Cetacea, Odontoceti, Physeteroidea) from the latest Miocene of Peru. *Comptes Rendu Palevol* 7: 361-369.

Lambert, O., Bianucci, G., Post, K., Muizon, C. de, Salas-Gismondi, R., Urbina, M., Reumer, J. 2010. The giant bite of a new raptorial sperm whale from the Miocene epoch of Peru. *Nature* 466: 105-108.

Lambert, O., Muizon, C. de. 2013. A new long-snouted species of the Miocene pontoporiid dolphin *Brachydelphis* and a review of the Mio-Pliocene marine mammal levels in the Sacaco Basin, Peru. *Journal of Vertebrate Paleontology* 33(3): 709-721.

Lambert, O., Muizon, C. de, Bianucci, G. 2013. The most basal beaked whale *Ninoziphius platyrostris* Muizon, 1983: clues on the evolutionary history of the family Ziphiidae (Cetacea, Odontoceti). *Zoological Journal of the Linnean Society* 167: 569-598.

Lindqvist, C., Bachmann, L., Anderson, L., Born, E.W., Arnason, U., Kovacs, K.M., Lydersen, C., Abramov, A.V., Wiig, O. 2009. The Laptev sea walrus *Odobenus rosmarus laptevi*: an enigma revisited. *Zoologica Scripta* 38: 113-127.

Marx, F., Kohno, N. 2016. A new Miocene baleen whale from the Peruvian desert. *Royal Society Open Science*, doi: 10.6084/m9.

McKee, J.W.A. 1994. Geology and vertebrate paleontology of the Tangahoe Formation, south Taranaki coast, New Zealand. *Geological Society Miscellaneous Publication* 80B: 63-91.

Mead, J.I., Spies, A.E., Sobolik, K.D. 2000. Skeleton of extinct North American sea mink. *Quaternary Research* 53: 247-262.

Morgan, G.S. 1994. Miocene and Pliocene marine mammal faunas from the Bone Valley Formation of central Florida. *Proceedings of the San Diego Society of Natural History* 29: 239-268.

Murakami, M., Shimada, C., Hikida, Y., Hirano, H. 2012a. A new basal porpoise, *Pterophocaena nishinoi* (Cetacea, Odontoceti, Delphinoidea), from the upper Miocene of Japan and its phylogenetic relationships. *Journal of Vertebrate Paleontology* 32: 1157-1171.

———. 2012b. Two new extinct basal phocoenids (Cetacea, Odontoceti, Delphinoidea), from the upper Miocene Koetoi Formation of Japan and their phylogenetic significance. *Journal of Vertebrate Paleontology* 32: 1172-1185.

Murakami, M., Shimada, C., Hikida, Y., Soeda, Y., Hiranov, H. 2014. *Eodelphis kabatensis*, a new name for the oldest true dolphin *Stenella kabatensis* Horikawa, 1977 (Cetacea, Odontoceti, Delphinidae) from the upper Miocene of Japan, and the phylogeny and paleobiogeography of Delphinoidea. *Journal of Vertebrate Paleontology* 34: 491-511.

Perrin, W.F., Brownell, R.L. Jr. 2001. Appendix 1 (of Annex U). update on the list of recognized species of cetaceans. *Journal of Cetacean Research and Management* 3(suppl.): 364-365.

Pyenson, N.D., Vélez-Juarbe, J., Gutstein, C.S., Little, H., Vigil, D., O'Dea, A. 2015. *Isthminia panamensis*, a new fossil inioid (Mammalia, Cetacea) from the Chagres Formation of Panama and the evolution of "river dolphins" in the Americas. *PeerJ* 3: e1227. doi: 10.7717.

Rosel, P.E., Dizon, A.E., Haygood, M.G. 1995. Variability of the mitochondrial control region in populations of the harbour porpoises, *Phocoena phocoena*, on interoceanic and regional scales. *Canadian Journal of Fisheries and Aquatic Sciences* 52: 1210-1219.

Scheel, D.M., Slater, G.J., Kolokotronis, S.O., Potter, C.W., Rotstein, D.S., Tsangaras, K., Greenwood, A.D., Helgen, K.M. 2014. Biogeography and taxonomy of extinct and endangered monk seals illuminated by ancient DNA and skull morphology. *ZooKeys* 409: 1-33.

Valenzuela-Toro, A.M., Pyenson, N.D., Gutstein, C.S, Suarez, M.E. 2016. A new dwarf seal from the late Neogene of South America and the evolution of pinnipeds in the southern hemisphere. *Papers in Paleontology* 2(1): 101-115.

Van der Feen, P.J. 1968. A fossil skull fragment of a walrus from the mouth of the River Scheldt (Netherlands). *Bijdragen tot de dierkunde* 38: 23-30.

Vélez-Juarbe, J., Domning, D.P. 2014a. Fossil Sirenia of the West Atlantic-Caribbean region IX. *Metaxytherium albifontanum*. *Journal of Vertebrate Paleontology* 34: 444-464.

———. 2014b. Fossil Sirenia of the West Atlantic and Caribbean region. X. *Priscosiren atlantica*, gen. et sp. nov. *Journal of Vertebrate Paleontology* 34: 951-964.

Vélez-Juarbe, J., Pyenson, N.D. 2012. *Bohaskaia monodontoides*, a new monodontid (Cetacea, Odontoceti, Delphinoidea) from the Pliocene of the western North Atlantic Ocean. *Journal of Vertebrate Paleontology* 32: 476-484.

Vélez-Juarbe, J., Domning, D.P., Pyenson, N.D. 2012. Iterative evolution of sympatric seacow (Dugongidae, Sirenia) assemblages during the past ~26 million years. *PLOS ONE* 7: e31294.

Vélez-Juarbe, J., Wood A.R., De Gracia C., Hendy, A.J.W. 2015. Evolutionary patterns among living and fossil kogiid sperm whales: evidence from the Neogene of Central America. *PLOS ONE* 10(4): e0123909. doi: 10.1371/journal .pone.

Wang, J.Y., Frasier, T.R., Yang, S.C., White, B.N. 2008. Detecting recent speciation events: The case of the finless porpoise (genus *Neophocaena*). *Heredity (Edinburgh)* 101(2): 145-155.

Whitmore, F. 1994. Neogene climatic change and the emergence of the modern whale fauna of the North Atl. Ocean. *Proceedings of the San Diego Society of Natural History* 29: 223-227.

Zhou, K., Zhou, M., Zhao, Z. 1984. First discovery of a Tertiary platanistoid fossil from Asia. *Scientific Reports of the Whales Research Institute* 35: 173-181.

Glossary

aDNA. Ancient DNA that has been recovered from archaeological or biological specimens.

Afrotheria. A mammalian clade with members that originate in Africa.

ambergris. A waxy substance that forms around squid beaks found in the intestine of some whales.

anagenesis. Evolution within a lineage in which one species evolves into another without splitting of the lineage.

"archaeocetes." A nonmonophyletic group of stem cetaceans.

avascular (osteo)necrosis. Death of bone tissue due to a lack of blood supply.

bathymetry. Measurement of the depth of bodies of water.

bottom-up. Refers to ecosystems controlled by nutrient availability.

bunodont. Having cheek teeth with low, rounded cusps.

bycatch. The proportion of a catch or harvest that includes nontargeted animals such as those taken incidentally through fishing.

Cetartiodactyla. A superorder of mammals that includes cetaceans and artiodactyls (even-toed ungulates).

characters. Heritable morphological, molecular, physiological, or behavioral attributes of organisms.

chemosynthesizers. Organisms such as bacteria that use chemical energy rather than photosynthesis to produce carbohydrates.

cladistics. A method of reconstructing the evolutionary history of organisms based on the shared possession of derived characters that provide evidence of a shared common ancestry.

cladogram. A branching diagram that depicts relationships among organisms based on the recency of common ancestry.

click. A broad-band frequency of sound of short duration.

convergence. Similarity between organisms that is not the result of inheritance.

countercurrent exchangers. Structures allowing an opposite flow of adjacent fluids that maximizes heat transfer rate.

crown group. Living members of a group and their common ancestors.

decompression syndrome. A serious condition that results when a diver breathes air under pressure and ascends too rapidly after spending time at depth. It occurs when nitrogen escapes into the blood, joints, and nerve tissue, causing pain, paralysis, and death unless treated by gradual decompression. Also known as "the bends."

denticles. Small tooth-like structures (e.g., cusps).

diastema. A space or gap between teeth.

digital acoustic tags. Small devices that measure the response of a wild marine mammal to sound.

digitigrade. Walking on the digits (fingers and toes).

diphyletic. Having two separate ancestries.

directional cranial asymmetry. In odontocetes, cranial bones and soft tissues on the right side are larger than those on the left side.

domoic acid. A toxin produced by various marine diatom species.

durophagous. Feeding on hard-shelled organisms such as shelled molluscs or crabs.

echolocation. The production of high-frequency sound and its reception by reflected echoes; used by toothed whales to navigate and locate prey.

encephalization quotient. A numerical comparison that considers brain size relative to body size.

endemic. Confined to a particular geographic region.

endocast. A mold that preserves the outline of a structure.

endothermy. Regulation of body temperature by muscular activity.

epochs. Subdivisions of geologic time.

finite element analysis. A method for predicting how a structure reacts to force, vibration, and fluid flow.

fluke. The horizontally flattened tail of cetaceans and sirenians.

geometric morphometrics. Analysis of shape using geometric coordinates.

ghost lineage. An evolutionary lineage that lacks a fossil record but the presence of which is inferred from the record of related taxa.

greenhouse gases. Compounds capable of absorbing radiation, thereby trapping heat in the atmosphere.

harmful algal blooms. Rapid increases in the population of algae in a water system. Also abbreviated as HAB.

heterodont. Having dentition in which teeth are differentiated into several types, such as incisors, canines, premolars, and molars, with different functions.

homodont. Having a dentition in which all teeth are very similar in form and function.

homologous characters. Characters in two taxa that were inherited from a common ancestor.

hypercarnivorous. Having an increased slicing component to the teeth.

hyperphalangy. An increase in number of digits, seen in whales.

ingroup. The group of organisms whose evolutionary relationships are the subject of investigation; see also **outgroup**.

kelp. A type of alga (seaweed).

keystone species. Species that play a critical role in an ecosystem.

krill. Shrimp-like crustaceans of the family Euphausiidae that form the primary food resource for baleen whales and certain pinnipeds (e.g., crabeater seal).

kya. Thousand years ago. Also abbreviated as KYA or ka.

lineage. An ancestor-descendant population.

lineage sorting. The process by which, following separation of two species, the ancestry of every gene converges to the overall phylogeny of the species.

lipids. A group of naturally occurring molecules including fats, waxes. and sterols.

Ma. Million years ago. Also abbreviated as MYA.

macroevolution. Evolution above the species level.

mass extinction. A large-scale event involving the extinction of many unrelated species.

mass strandings. Three or more individuals of the same species that intentionally swim ashore or are unintentionally trapped ashore by waves or receding tides.

melon. Fat-filled structure on the forehead of odontocetes that functions to focus sound.

monophyletic. Refers to a group that consists of a common ancestor and all of its descendants.

myoglobin. Oxygen-binding protein found in muscle cells.

osteochondrosis. Disease that affects the growth of bones.

osteosclerotic. Refers to compact bone.

outgroup. Taxon or group of organisms that is closely related but outside the group whose relationships are the subject of investigation; see also **ingroup**.

outgroup comparison. Procedure for determining the polarity (ancestral or derived condition) of a character; assumes that the character found in the outgroup is the ancestral condition for the group in question (the ingroup).

pachyostotic. Refers to thick, dense bone.

paedomorphism. A change in developmental timing such that adults retain juvenile characteristics.

Paenungulata. Clade of mammals that unites proboscideans, sirenians, and hyracoids.

panmictic. Describing a mating strategy with a random mating pattern.

paraphyletic. Refers to a group that consists of the common ancestor but not all the descendants.

Paratethys. A former large, shallow sea that extended across central Europe and eastern Asia.

passive selection. Trend in evolution that is the result of diversification of a clade and subsequent filling of available morphospace; contrasts with active selection, which results in a directional change in a lineage.

peramorphosis. An increase in developmental timing in which maturity is delayed while the development of the adult is extended.

phonic lips. Soft tissue structures located in the nasal passages that are hypothesized as a mechanism of sound production.

phylogeny. Evolutionary history of a group of organisms.

pinnipedimorphs. A clade (Pinnipedimorpha) that includes living pinnipeds and their extinct fossil relatives.

piscivorous. Feeding mostly on fish.

plantigrady. Type of standing posture in which the sole of the foot is on the ground.

plate tectonics. The concept that the earth's surface is organized into large mobile blocks of crustal material.

polarity. The direction of character state change (ancestral vs. derived).

polydont. Having an increased number of teeth.

polygynous. Describing a mating strategy in which one male mates with more than one female during a breeding season.

polyphyletic. Refers to a type of nonmonophyletic group in which the group is characterized by a convergently evolved character.

polytomy. A pattern of unresolved relationships among taxa.

principal components analysis. A statistical procedure that converts a set of observations of possibly correlated variables into a set of values of linearly uncorrelated variables (principal components).

pseudogenes. Relatives of genes that are nonfunctional and have lost their ability to code a protein.

rhizomes. Modified underground stems of plants.

scientific name. Unique name of a species, consisting of a genus and species name (specific epithet).

sexual dimorphism. External differences between males and females of a particular species.

sexual selection. A special type of natural selection in which the sexes acquire distinct forms either because the members of one sex choose mates with particular features or because, in the competition for mates among members of one sex, only those with certain traits succeed.

sister taxa. A pair of species that are each other's closest relative.

stem group. Organisms that fall outside the crown group.

sympatric. Refers to two populations or species that exist in the same place.

synapomorphies. Shared derived characters, used as evidence of a shared common ancestry between taxa.

stratigraphy. Branch of geology that studies rock layers (strata) and their interpretation.

taphonomy. Study of processes that affect an organism after death.

taxon. A particular taxonomic group at a given rank; plural, *taxa.*

Tethys Sea. A former shallow sea that extended in an area now occupied by India and Pakistan.

Tethytheria. A monophyletic group that includes proboscideans (elephants), sirenians, and extinct desmostylians.

teuthophagous. Feeding mostly on cephalopods (e.g., squid).

time scale. A system that relates stratigraphy to time.

tooth isotopes. Isotope signatures (distribution of isotopes) that can be measured in teeth and reflect an animal's diet.

top-down. Refers to ecosystems controlled by predators.

upwelling. Area off continental margins where the circulation patterns bring nutrient-rich water to the surface.

vouchered specimen. Reference specimen together with details of its locality and collecting information necessary to prevent misidentification.

whale falls. Organisms that evolve around the carcasses of whales on the sea bottom.

Ypb. Years before the present.

zooplankton. Animal plankton found in the upper portions of the ocean.

References

Chapter 1. Setting the Stage

Berta, A., Sumich, J.L., Kovacs, K.M. 2015. *Marine Mammals: Evolutionary Biology*, 3rd ed. Elsevier, San Diego, CA.

Kuntner, M., May-Collado, L.J., Agnarsson, I. 2010. Phylogeny and conservation priorities of afrotherian mammals (Afrotheria, Mammalia). *Zoologica Scripta* 40: 1-15.

Marx, F. 2009. Marine mammals through time: when less is more in studying paleodiversity. *Proceedings of the Royal Society B* 276: 887-892.

Mirceta, S., Signore, A.V., Burns, J.M., Cossins, A.R., Campbell, K.L., Berenbrink, M. 2013. Evolution of mammalian diving capacity traced by myoglobin net surface charge. *Science* 340: 1234192.

Nowacek, D.P., Johnson, M.P., Tyack, P.L., Shorter, K.A., McLellan, W.A., Pabst, A. 2001. Buoyant balaenids: the ups and downs of buoyancy in right whales. *Proceedings of the Royal Society B* 268: 1811-1816.

Pyenson, N.D., Gutstein, C.S., Parham, J.F., Le Roux, J.P., Chavarría, C.C., Little, H., Metallo, A., Rossi, V., Valenzuela-Toro, A.M., Vélez-Juarbe, J., Santelli, C.M., Rubilar Rogers, D., Cozzuol, M.A., Suárez, M.E. 2014. Repeated mass strandings of Miocene marine mammals from Atacama region of Chile point to sudden death at sea. *Proceedings of the Royal Society B* 281(1781). doi: 10.1098/rspb.2013.3316.

Shubin, N. 2001. Quoted in *Whale Evolution*. WGBH Educational Foundation and Clear Blue Sky Productions. www.pbs.org/wgbh/evolution/library/03/4/l_034_05.html.

Thewissen, J.G.M. 2015. *The Walking Whales: From Land to Water in 8 Million Years*. University of California Press, Berkeley, CA.

Thewissen, J.G.M., Bajpai, S. 2009. New skeletal material of *Andrewsiphius* and *Kutchicetus*, two Eocene cetaceans from India. *Journal of Paleontology* 83: 635-663.

Thewissen, J.G.M., Cooper, L.N., Clementz, M.T., Bajpai, S., Tiwari, B.N. 2007. Whales originated from aquatic artiodactyls in the Eocene epoch of India. *Nature* 450: 1190-1194.

Uhen, M.D., Pyenson N.D. 2007. Diversity estimates, biases, and historiographic effects: resolving cetacean diversity in the Tertiary. *Palaeontologica Electronica* 10(2): 11A.

Chapter 2. The Oldest Marine Mammals

Andrews, C.W. 1902a. Dr. C. W. Andrews on fossil vertebrates from Upper Egypt. *Proceedings of the Zoological Society of London* 1902: 228-230.

———. 1902b. Preliminary note on some recently discovered extinct vertebrates from Egypt, part III. *Geological Magazine*, Decade IV, 9: 291-295.

———. 1904. Further notes on the mammals of the Eocene of Egypt, part III. *Geological Magazine*, Decade V, 1: 211-215.

Bebej, R.M., ul-Haq, M., Zalmout, I.S., Gingerich, P.D. 2012. Morphology and function of the vertebral column in *Remingtonocetus domandensis* (Mammalia, Cetacea) from the middle Eocene Domanda Formation of Pakistan. *Journal of Mammalian Evolution* 19: 77-104.

Bebej, R.M., Zalmout, I.S., Abed El-Aziz, A.A., Antar, M.S.M., Gingerich, P.D. 2016. First remingtonocetid archaeocete (Mammalia, Cetacea) from the middle Eocene of Egypt with implications for biogeography and locomotion in early cetacean evolution. *Journal of Paleontology*, doi: 10.1017/jpa.2015.57.

Benoit, J., Adnet, S., El Mabrouk, E., Khayati, H., Ali, M.B.H., Marivaux, L., Merzeraud, G., Merigeaud, S., Vianey-Liaud, M., Tabuce, R. 2013. Cranial remain from Tunisia provides new clues for the origin and evolution of Sirenia (Mammalia, Afrotheria) in Africa. *PLOS ONE* 8: e54307.

Bianucci, G., Gingerich, P.D. 2011. *Aegyptocetus tarfa*, n. gen. et sp. (Mammalia, Cetacea), from the middle Eocene of Egypt: clinorhynchy, olfaction, and hearing in a protocetid whale. *Journal of Vertebrate Paleontology* 31: 1173-1188.

Buchholtz, E.A. 2007. Modular evolution of the cetacean vertebral column. *Evolution and Development* 9: 278-289.

Churchill, M., Martinez-Caceres, M., Muizon, C. de, Mnieckowski, J., Geisler, J.H. 2016. The origin of high-frequency hearing in whales. *Current Biology* 26: 1-6.

Cooper, L.N., Hieronymus, T.L., Vinyard, C., Bajpai, S.,

Thewissen, J.G.M. 2014. Applications for constrained ordination: reconstructing feeding behaviors in fossil Remingtonocetinae (Cetacea: Mammalia). In: Hebbree, D.I., Platte, B.F., Smith, J.J. (eds), *Experimental Approaches to Understanding Fossil Organisms, Topics in Geobiology*, vol. 41. Springer Science + Business Media, Dordrecht, Netherlands, pp. 89-107.

Domning, D.P. 2000. The readaptation of Eocene sirenians to life in the water. *Historical Biology* 14: 115-119.

Ekdale, E.G., Racicot, R.A. 2015. Anatomical evidence for low frequency sensitivity in an archaeocete whale: comparison of the inner ear of *Zygorhiza kochii* with that of crown Mysticeti. *Journal of Anatomy* 226(1): 22-39.

Fahlke, J.M., Hampe, O. 2015. Cranial symmetry in baleen whales (Cetacea, Mysticeti) and the occurrence of cranial asymmetry throughout cetacean evolution. *Naturwissenschaften* 102(9-10): 1309.

Fahlke, J.M., Gingerich, P.D., Welsh, R.C., Wood, A.R. 2011. Cranial asymmetry in Eocene archaeocete whales and the evolution of directional hearing. *Proceedings of the National Academy of Sciences USA* 408: 14,545-14,548.

Fahlke, J.M., Bastl, K.A., Semprebonn, G.M., Gingerich, P.D. 2013. Paleoecology of archaeocete whales throughout the Eocene: dietary adaptations revealed by microwear analysis. *Palaeogeography, Palaeoclimatology, and Palaeoecology* 386: 690-701.

Fordyce, R.E. 2009. Cetacean evolution. In: Perrin, W.F., Wursig, B., Thewissen, J.G.M. (eds), *Encyclopedia of Marine Mammals*, 2nd ed. Elsevier, San Diego, CA, pp. 201-207.

Gatesy, J., Geisler, J.H., Chang, J., Buell, C., Berta, A., Meredith, R.W., Springer, M.S., McGowen, M.R. 2013. A phylogenetic blueprint for a modern whale. *Molecular Phylogenetics and Evolution* 66: 479-506.

Gingerich, P.D. 1992. Marine mammals (Cetacea and Sirenia) from the Eocene of Gebel Mokattam and Fayum, Egypt: stratigraphy, age and paleoenvironments. *Papers on Paleontology, University of Michigan*, no. 30.

———. 2015a. Evolution of whales from land to sea. In: Dial, K.P., Shubin, N., Brainerd, E.L. (eds), *Major Transformations in Vertebrate Evolution*. University of Chicago Press, Chicago, IL, pp. 239-256.

———. 2015b. New partial skeleton and relative brain size in the late Eocene archaeocete *Zygorhiza kochii* (Mammalia, Cetacea) from the Pachuta marl of Alabama, with a note on contemporaneous *Pontogeneus brachyspondylus*. *Contributions from the Museum of Paleontology, University of Michigan* 32(10): 161-188.

———. 2016. Body weight and relative brain size (encephali-zation) in Eocene Archaeoceti (Cetacea). *Journal of Mammalian Evolution* 23(1): 17-31.

Gingerich, P.D., Russell, D.E. 1981. *Pakicetus inachus*, a new archaeocete (Mammalia: Cetacea). *Contributions from the Museum of Paleontology, University of Michigan* 25: 235-246.

Gingerich, P.D., Domning, D.P., Blane, C.E., Uhen, M.D. 1994. Cranial morphology of *Protosiren fraasi* (Mammalia, Sirenia) from the middle Eocene of Egypt: a new study using computed tomography. *Contributions from the Museum of Paleontology, University of Michigan* 29: 41-67.

Godfrey, S.J. 2013. On the olfactory apparatus in the Miocene odontocete *Squalodon* sp. (Squalodontidae). *Comptes Rendus Palevol* 12: 519-530.

Godfrey, S.J., Geisler, J., Fitzgerald, E. 2012. On the olfactory anatomy in an archaic whale (Protocetidae, Cetacea) and the minke whale *Balaenoptera acutorostrata* (Balaenopteridae, Cetacea). *Anatomical Record* 296: 257-272.

Gol'din, P., Zvonok, E. 2013. *Basilotritus uheni*, a new cetacean (Cetacea, Basilosauridae) from the late Middle Eocene of eastern Europe. *Journal of Paleontology* 87: 254-268.

Harlan, R. 1834. Notice of fossil bones found in the Tertiary formation of the state of Louisiana. *Transactions of the American Philosophical Society, Philadelphia* 4(12): 397-402.

Houssaye, A., Tafforeau, P., Muizon, C. de, Gingerich, P.D. 2015. Transition of Eocene whales from land to sea: evidence from bone microstructure. *PLOS ONE* 10(2): e0118409.

Kellogg, R. 1936. A review of the Archaeoceti. *Carnegie Institute of Washington Special Publication* 482: 1-366.

Kishida, T., Thewissen, J.G.M., Hayakawa, T., Imai, H., Agata, K. 2015. Aquatic adaptation and the evolution of smell and taste in whales. *Zoological Letters* 1: 9. doi: 10.1186/s40851-014-0002-z.

Newsome, S.D., Clementz, M.T., Koch, P.L. 2010. Using stable isotope biogeochemistry to study marine mammal ecology. *Marine Mammal Science* 26(3): 509-573.

Owen, R. 1839. Observations on the teeth of the *Zeuglodon, Basilosaurus* of Dr. Harlan. *Proceedings of the Geological Society of London* 1839: 24-28.

———. 1855. On the skull of a mammal (*Prorastomus sirenoides*, Owen), from the island of Jamaica. *Quarterly Journal of the Geological Society of London* 31: 559-567.

Ray, J. 1693. *Synopsis Methodica Animalium Quadrupedum et Serpentini Generis*. Robert Southwell, London.

Sniveley, E., Fahlke, J.M., Welsh, R.C. 2015. Bone-breaking bite force of *Basilosaurus isis* (Mammalia, Cetacea) from the late Eocene of Egypt estimated by finite element analysis. *PLOS ONE* 10(2): e0118380.

Stromer, E. 1908. *Die Archaeoceti des agyptischen Eozans.* W. Braumuller, Austria.

Thewissen, J.G.M. 2015. *The Walking Whales: From Land to Water in 8 Million years.* University of California Press, Berkeley, CA.

Thewissen, J.G.M., Williams, E.M. 2002. The early radiations of Cetacea (Mammalia): evolutionary pattern and developmental correlations. *Annual Reviews of Ecology and Systematics* 33: 73-90.

Thewissen, J.G.M., Hussain, S.T., Arif, M. 1994. Fossil evidence for the origin of aquatic locomotion in archaeocete whales. *Science* 263: 201-212.

Uhen, M.D. 2004. *Form, Function, and Anatomy of* Dorudon atrox *(Mammalia, Cetacea): An Archaeocete from the Middle to Late Eocene of Egypt*, University of Michigan, Papers on Paleontology, no. 34. University of Michigan, Ann Arbor, MI.

———. 2007. Evolution of marine mammals: back to the sea after 300 million years. *Anatomical Record* 290: 514-522.

Vélez-Juarbe, J., Domning, D.P. 2015. Fossil Sirenia of the West Atlantic and Caribbean region. XI: *Callistosiren boriquensis*, gen. et sp. nov. *Journal of Vertebrate Paleontology* 35(1): e885034.

West, R.M. 1980. Middle Eocene large mammal assemblage with Tethyan affinities, Ganda Kas region, Pakistan. *Journal of Paleontology* 54: 508-533.

Chapter 3. Later-Diverging Whales

Aguirre-Fernández, G., Fordyce, R.E. 2014. *Papahu taitapu*, gen. et sp. nov., an early Miocene stem odontocete (Cetacea) from New Zealand. *Journal of Vertebrate Paleontology* 34: 195-210.

Armfield, B., Zheng, Z., Bajpai, S., Vinyard, C.J., Thewissen, J.G.M. 2013. Development and evolution of the unique cetacean dentition. *PeerJ* 1: e24. http://dx.doi.org/10.7717/peerj.24.

Bajpai, S., Thewissen, J.G.M., Conley, R.W. 2011. Cranial anatomy of middle Eocene *Remingtonocetus* (Cetacea, Mammalia) from Kutch, India. *Journal of Paleontology* 85: 703-718.

Baldanza, A., Bizzarri, R., Famiani, F., Monaco, P., Pellegrino, R., Sassi, P. 2013. Enigmatic, biogenically induced structures in Pleistocene marine deposits: a first record of fossil ambergris. *Geology* 41(10): 1075-1078.

Barnes, L.G. 1978. A review of *Lophocetus* and *Liolithax* and their relationships to the delphinoid family Kentriodontidae (Cetacea: Odontoceti). *Natural History Museum of Los Angeles County Science Bulletin* 28: 1-35.

Beatty, B.L., Rothschild, B.M. 2008. Decompression syndrome and the evolution of deep diving physiology in cetaceans. *Naturwissenschaften* 95: 793-801.

Beneden, P.J. van. 1872. Les baleines fossiles d'anvers. *Bulletin de l'Academie des Sciences de Belgique* 34(2): 6-20.

Berta, A. (ed). 2015. *Whales, Dolphins, and Porpoises.* University of Chicago Press, Chicago, IL.

Berta, A., Sumich, J.L., Kovacs, K.M. 2015. *Marine Mammals: Evolutionary Biology.* Elsevier, San Diego, CA.

Berta, A., Lanzetti, A., Ekdale, E.G., Deméré, T.A. 2016. From teeth to baleen and raptorial to bulk filter feeding in mysticete cetaceans: the role of paleontologic, genetic, and geochemical data in feeding evolution and ecology. *Integrative and Comparative Biology* 56(6): 1271-1284.

Bianucci, G. 1996. The Odontoceti (Mammalia, Cetacea) from Italian Pliocene: systematics and phylogenesis of Delphinidae. *Palaeontographia Italica* 83: 73-167.

———. 2013. *Septidelphis morii*, n. gen. et sp., from the Pliocene of Italy: new evidence of the explosive radiation of true dolphins (Odontoceti, Delphinidae). *Journal of Vertebrate Paleontology* 33(3): 722-740.

Bianucci, G., Landini, W. 2002. Change in diversity, ecological significance and biogeographical relationships of the Mediterranean Miocene toothed whale fauna. *Geobios*: 19-28.

Bianucci, G., Lambert, O., Post, K. 2007. A high diversity in fossil beaked whales (Odontoceti, Ziphiidae) recovered by trawling from the sea floor off South Africa. *Geodiversitas* 29: 561-618.

Bianucci, G., Lambert, O., Salas-Gismondi, R., Tejada, J., Pujos, F., Urbina, M., Antoine, P.O. 2013a. A Miocene relative of the Ganges River dolphin (Odontoceti, Platanistidae) from the Amazon Basin. *Journal of Vertebrate Paleontology* 33(3): 741-745.

Bianucci, G., Mijan, I., Lambert, O., Post, K., Mateus, O. 2013b. Bizarre fossil beaked whales (Odontoceti, Ziphiidae) fished from the Atlantic Ocean floor off the Iberian Peninsula. *Geodiversitas* 35(1): 105-152.

Bianucci, G., Di Celma, C., Urbina, M., Lambert, O. 2016. New beaked whales from the late Miocene of Peru and evidence of convergent evolution in stem and crown Ziphiidae (Cetacea, Odontoceti). *PeerJ* 4: e2479. doi: 10.7717/peerj.2479.

Bisconti, M. 2005. Taxonomic revision and phylogenetic relationships of the rorqual-like mysticete from the Pliocene of Mount Pulgnasco. northern Italy (Mammalia, Cetacea, Mysticeti). *Palaeontographia Italica* 91: 85-108.

———. 2006. *Titanocetus,* a new baleen whale from the middle Miocene of northern Italy (Mammalia, Cetacea, Mysticeti). *Journal of Vertebrate Paleontology* 26(2): 344-354.

———. 2007. A new basal balaenopterid whale from the Pliocene of northern Italy. *Palaeontology* 50(5): 1103-1122.

———. 2012. Comparative osteology and phylogenetic relationships of *Miocaperea pulchra,* the first fossil pygmy right whale genus and species (Cetacea, Mysticeti, Neobalaenidae). *Zoological Journal of the Linnean Society* 166: 876-911.

———. 2015. Anatomy of a new cetotheriid genus and species from the Miocene of Herentals, Belgium, and the phylogenetic and palaeobiogeographical relationships of Cetotheriidae s.s. (Mammalia, Cetacea, Mysticeti). *Systematic Paleontology* 13(5): 377-395.

Bisconti, M., Bosselaers, M. 2016. *Fragilicetus velponi:* a new mysticete genus and species and its implications for the origin of Balaenopteridae (Mammalia, Cetacea, Mysticeti). *Zoological Journal of the Linnean Society* 177: 450-474.

Bisconti, M., Lambert, O., Bosselaers, M. 2013. Taxonomic revision of *Isocetus depauwi* (Mammalia, Cetacea, Mysticeti) and the phylogenetic relationships of archaic "cetothere" mysticetes. *Palaeontology* 56(1): 95-127.

Boersma, A.T., Pyenson, N.D. 2015. *Albicetus oxymycterus,* a new generic name and redescription of a basal physeteroid (Mammalia, Cetacea) from the Miocene of California, and the evolution of body size in sperm whales. *PLOS ONE* 10(12): e0135551.

———. 2016. *Arktocara yakataga,* a new fossil odontocete (Mammalia, Cetacea) from the Oligocene of Alaska and the antiquity of Platanistoidea. *Peerj,* doi: 10.7717/peerj .2321.

Boessenecker, R.W. 2013. A new marine vertebrate assemblage from the Late Neogene Purisima Formation in Central California, part II: pinnipeds and cetaceans. *Geodiversitas* 35(4): 815-940.

Boessenecker, R.W., Fordyce, R.E. 2015a. Anatomy, feeding ecology, and ontogeny of a transitional baleen whale: a new genus and species of Eomysticetidae (Mammalia: Cetacea) from the Oligocene of New Zealand. *PeerJ,* doi: 10.7717/peerj.1129.

———. 2015b. A new eomysticetid (Mammalia: Cetacea) from the Late Oligocene of New Zealand and a re-evaluation of *"Mauicetus" waitakiensis. Papers in Palaeontology* 1: 107-140. doi: 10.1002/spp2.1005.

———. 2015c. A new genus and species of eomysticetid (Cetacea: Mysticeti) and a reinterpretation of *"Mauicetus lophocephalus"* Marples, 1956: transitional baleen whales from the upper Oligocene of New Zealand. *Zoological Journal of the Linnean Society* 175(3): 607-660.

———. 2016. A new eomysticetid from the Oligocene Kokoamu Greensand of New Zealand and a review of the Eomysticetidae (Mammalia, Cetacea). *Journal of Systematic Paleontology,* doi: 10.1080/14772019.2016.1191045.

Boessenecker, R.W., Poust, A.W. 2015. Freshwater occurrence of the extinct dolphin *Parapontoporia* (Cetacea: Lipotidae) from the upper Pliocene nonmarine Tulare Formation of California. *Palaeontology* 58(3): 489-496.

Boessenecker, R.W., Perry, F.A., Geisler, J.H. 2015. Globicephaline whales from the Mio-Pliocene Purisima Formation of central California, USA. *Acta Palaeontologica Polonica* 60(1): 113-122.

Buono, M.R., Dozo, M.T., Marx, F.G., Fordyce, R.E. 2014a. A late Miocene potential Neobalaenine mandible from Argentina sheds light on the origins of the living pygmy right whale. *Palaeontologica Polonica* 59(4): 787-793.

Buono, M.R., Fernández, M.S., Cozzuol, M.A. 2014b. Miocene balaenids (Cetacea: Mysticeti: Balaenidae) from Patagonia (Argentina) and their implication for early evolution of right whales. Abstract, 4th International Congress Mendoza, Argentina, p. 662.

Churchill, M., Martinez-Caceres, M., Muizon, C. de, Mnieckowski, J., Geisler, J.H. 2016. The origin of high-frequency hearing in whales. *Current Biology* 26: 1-6.

Clementz, M.T., Fordyce, R.E., Peek, S.L., Fox, D.L. 2014. Ancient marine isoscapes and isotopic evidence of bulk-feeding by Oligocene cetaceans, *Palaeogeography, Palaeoclimatology, and Palaeoecology* 400: 28-40.

Collareta, A., Landini, W., Lambert, O., Post, K., Tinelli, C., Di Celma, C., Panetta, D., Tripoli, M., Salvadori, P., Caramella, D., Marchi, D., Urbina, M., Bianucci, G. 2015. Piscivory in a Miocene Cetotheriidae of Peru: first-record of fossilized stomach content for an extinct baleen-bearing whale. *Science Nature,* doi: 10.1007/s00114 -015-1319-y.

Cooper, L.N., Berta, A., Dawson, S.D., Reidenberg, J.S. 2007. Evolution of hyperphalangy and digit reduction in the cetacean manus. *Anatomical Record* 209: 654-672.

Cozzuol, M. 2010. Fossil record and the evolutionary history of Iniodea, In: Ruiz-Garcia, M., Shostell, J. (eds), *Biology, Evolution, and Conservation of River Dolphins.* Nova Science, Hauppauge, NY, pp. 193-217.

Cranford, T.W., Krysl, P., Hildebrand, J.A. 2008a. Acoustic pathway revealed: simulated sound transmission and reception in Cuvier's beaked whale *(Ziphius cavirostris). Bioinspiration and Biomimetics* 3: 1-10.

Cranford, T.W., McKenna, M.F., Soldevilla, M.S., Wiggins, S.M., Goldbogen, J.A., Shadwick, R.E., Krysl, P., St Leger, J.A., Hildebrand, J.A. 2008b. Anatomic geometry of sound transmission and reception in Cuvier's beaked whale (*Ziphius cavirostris*). *Anatomical Record* 291: 353-378.

Cranford, T.W., Krysl, P., Amundin, M. 2010. A new acoustic portal into the odontocete ear and vibrational analysis of the tympanoperiotic complex. *PLOS ONE* 5: e11927.

Deméré, T.A., McGowen, M.R., Berta, A., Gatesy, J. 2008. Morphological and molecular evidence for a stepwise evolutionary transition from teeth to baleen in mysticete whales. *Systematic Biology* 57: 15-37.

Dines, J.P., Otárola-Castillo, E., Ralph, P., Alas, J., Daley, T., Smith, A.D., Dean, M.D. 2014. Sexual selection targets cetacean pelvic bones. *Evolution* 68(11): 3296-3306.

Ekdale, E.G., Demere, T.A., Berta, A. 2015. Vascularization of the gray whale plate (Cetacea, Mysticeti, *Eschrichtius robustus*): soft tissue evidence for an alveolar source of blood to baleen. *Anatomical Record* 298: 691-702.

El Adli, J.J., Deméré, T.A., Boessenecker, R.W. 2014. *Herpetocetus morrowi* (Cetacea: Mysticeti), a new species of diminutive baleen whale from the upper Pliocene (Piacenzian) of California, USA, with observations on the evolution and relationships of the Cetotheriidae. *Zoological Journal of the Linnean Society* 170: 400-466.

Fahlke, J., Hampe, O. 2015. Cranial symmetry in baleen whales (Cetacea, Mysticeti) and the occurrence of cranial asymmetry throughout cetacean evolution. *Naturwissenschaften* 102(9-10): 1309.

Fitzgerald, E.M.G. 2006. A bizarre new toothed mysticete (Cetacea) from Australia and the early evolution of baleen whales. *Proceedings of the Royal Society B* 273: 2955-2963.

———. 2010. The morphology and systematics of *Mammalodon colliveri* (Cetacea: Mysticeti), a toothed mysticete from the Oligocene of Australia. *Zoological Journal of the Linnean Society* 158: 367-476.

———. 2012a. Archaeocete jaws in a baleen whale. *Biology Letters* 8(1): 94-96.

———. 2012b. Possible neobalaenid from the Miocene of Australia implies a long evolutionary history for the pygmy right whale *Caperea marginata* (Cetacea, Mysticeti). *Journal of Vertebrate Paleontology* 32: 976-980.

Fordyce, R.E. 1994. *Waipatia maerewhenua*, new genus and new species (Waipatiidae, new family), an archaic Late Oligocene dolphin (Cetacea: Odontoceti: Platanistoidea) from New Zealand. In: Berta, A., Deméré, T. (eds), *Contributions in Marine Mammal Paleontology Honoring Frank C. Whitmore, Jr.*, *Proceedings of the San Diego Society of Natural History*, vol. 29. San Diego Museum of Natural History San Diego, CA, pp. 147-176.

———. 2009. Cetacean evolution. In: Perrin, W.F., Wursig, B., Thewissen, J.G.M. (eds), *Encyclopedia of Marine Mammals*, 2nd ed. Elsevier, San Diego, CA, pp. 201-207.

Fordyce, R.E., Marx, F.G. 2013. The pygmy right whale *Capera marginata*: the last of the cetotheres. *Proceedings of the Royal Society B* 280(1753). doi: 10.1098/rspb.2012.2645.

———. 2016. Mysticetes baring their teeth: a new fossil whale, *Mammalodon hakataramea*, from the Southwest Pacific. *Memoirs of the Museum Victoria* 74: 107-116.

Friedman, M., Shimada, K., Martin, L.D., Everhart, M.J., Liston, J., Maltese, A., Triebold, M. 2010. 100-million year dynasty of giant planktivorous bony fishes in the Mesozoic seas. *Science* 327: 990-993.

Galatius, A., Berta, A., Frandsen, M.S., Goodall, R.N.P. 2011. Interspecific variation on ontogeny and skull shape among porpoises (Phocoenidae). *Journal of Morphology* 272: 136-148.

Gatesy, J., Geisler, J.H., Chang, J., Buell, C., Berta, A., Meredith, R.W., Springer, M.S., McGowen, M.R. 2013. A phylogenetic blueprint for a modern whale. *Molecular Phylogenetics and Evolution* 66: 479-506.

Geisler, J.H., McGowen, M.R., Yang, G., Gatesy, J. 2011. A supermatrix analysis of genomic, morphological and paleontological data from crown Cetacea. *BMC Evolutionary Biology* 11: 1-33.

Geisler, J.H., Godfrey, S.J., Lambert, O. 2012. A new genus and species of late Miocene inioid (Cetacea, Odontoceti) from the Meherrin River, North Carolina, U.S.A. *Journal of Vertebrate Paleontology* 32(1): 198-211.

Geisler, J.H., Colbert, M.W., Carew, J.L. 2014. A new fossil species supports an early origin for toothed whale echolocation. *Nature* 508: 383-386.

Gibson, M.L., Geisler, J.H. 2009. A new Pliocene dolphin (Cetacea: Pontoporiidae), from the Lee Creek Mine, North Carolina. *Journal of Vertebrate Paleontology* 29(3): 966-971.

Gingerich, P.D. 2015. New partial skeleton and relative brain size in the late Eocene archaeocete *Zygorhiza kochii* (Mammalia, Cetacea) from the Pachuta marl of Alabama, with a note on contemporaneous *Pontogeneus brachyspondylus*. *Contributions from the Museum of Paleontology, University of Michigan* 32(10): 161-188.

———. 2016. Body weight and relative brain size (encephalization) in Eocene Archaeoceti (Cetacea). *Journal of Mammalian Evolution* 23(1): 17-31.

Godfrey, S.J., Barnes, L.G. 2008. A new genus and species of late Miocene pontoporiid dolphin (Cetacea: Odontoceti) from the St. Marys Formation in Maryland. *Journal of Vertebrate Paleontology* 28(2): 520-528.

Godfrey, S.J., Uhen, M.D., Osborne J.E., Edwards, L.E. 2016. A new specimen of *Agorophius pygmaeus* (Agorophiidae, Odontoceti, Cetacea) from the early Oligocene Ashley Formation of South Carolina, USA. *Journal of Paleontology* 90: 154-169.

Gol'din, P. 2014. "Antlers inside": are the skull structures of beaked whales (Cetacea: Ziphiidae) used for echoic imaging and visual display? *Biological Journal of the Linnean Society* 113: 510-515.

Gol'din, P., Startsev, D. 2014. *Brandtocetus*, a new genus of baleen whales (Cetacea, Cetotheriidae) from the late Miocene of Crimea, Ukraine. *Journal of Vertebrate Paleontology* 34(2): 419-433.

Gol'din, P., Steeman, M.E. 2015. From problem taxa to problem solver: a new Miocene family, Tranatocetidae, brings perspective on baleen whale evolution. *PLOS ONE* 10(9): e0135500. doi: 10.1371/journal.pone.0135500.

Hrbek, T., da Silva, V.M.F., Dutra, N., Gravena, W., Martin, A.R., Farias, P.I. 2014. A new species of river dolphin from Brazil or: how little do we know our biodiversity. *PLOS ONE* 9(1): e83623.

Hussain, A., Delsuc, F., Ropiquet, A., Hammer, C., Jansen Van Vuuren, B., Mathee, C., Ruiz-Garcia, M., Catzeflis, F., Areskoug, V., Nguyen, T.T., Couloux, A. 2012. Pattern and timing of diversification of Cetartiodactyla (Mammalia, Laurasiatheria), as revealed by a comprehensive analysis of mitochondrial genomes. *Comptes Rendus Biologies* 335: 32-50.

Kellogg, R. 1923. Description of an apparently new toothed cetacean from South Carolina. *Smithsonian Miscellaneous Collections* 76(7): 1-7.

———. 1924. Description of a new species of whalebone whale from the Calvert Cliffs, Maryland. *Proceedings of the US National Museum* 63: 1-14.

Lambert, O., Muizon, C. de. 2013. A new long-snouted species of the Miocene pontoporiid dolphin *Brachydelphis* and a review of the Mio-Pliocene marine mammal levels in the Sacaco Basin, Peru. *Journal of Vertebrate Paleontology* 33(3): 709-721.

Lambert, O., Bianucci, G., Post, K., Muizon, C. de, Salas-Gismondi, R., Urbina, M., Reumer, J. 2010. The giant bite of a new raptorial sperm whale from the Miocene epoch of Peru. *Nature* 466: 105-108.

Lambert, O., Muizon, C. de, Bianucci, G. 2013. The most basal beaked whale *Ninoziphius platyrostris* Muizon, 1983:

clues on the evolutionary history of the family Ziphiidae (Cetacea: Odontoceti). *Zoological Journal of the Linnean Society* 167: 569-598.

Lambert, O., Bianucci, G., Beatty, B.L. 2014a. Bony outgrowths on the jaws of an extinct sperm whale support macroraptorial feeding in several stem physeteroids. *Naturwissenschaften* 101: 517-521.

Lambert, O., Bianucci, G., Urbina, M. 2014b. *Huaridelphis raimondii*, a new early Miocene Squalodelphinidae (Cetacea, Odontoceti) from the Chilcatay Formation, Peru. *Journal of Vertebrate Paleontology* 34(5): 987-1004.

Lambert, O., Collareta, A., Landini, W., Post, K., Ramassamy, B., Di Celma, C., Urbina, M., Bianucci, G. 2015a. No deep diving: evidence of predation on epipelagic fish for a stem beaked whale from the Late Miocene of Peru. *Proceedings of the Royal Society B* 282(1815). doi: 10.1098/rspb.2015.1530.

Lambert O., Muizon, C. de, Bianucci, G. 2015b. A new archaic homodont toothed cetacean (Mammalia, Cetacea, Odontoceti) from the Early Miocene of Peru. *Geodiversitas* 37(1): 79-108.

Lambert, O., Bianucci, G., Muizon, C. de. 2016. Macroraptorial sperm whales (Cetacea, Odontoceti, Physeteroidea) from the Miocene of Peru. *Zoological Journal of the Linnean Society*, doi: 10.111/zoj.12456.

Lambert, O., Bianucci, G., Urbina, M., Geisler, J. 2017. A new inioid from the Miocene of Peru and the origin of modern dolphin and porpoise families. *Zoological Journal of the Linnean Society* 179: 919-946.

Lindberg, D.R., Pyenson, N.D. 2007. Things that go bump in the night: evolutionary interactions between cephalopods and cetaceans in the Tertiary. *Lethaia* 40(4): 335-342.

Loch, C., Kieser, J.A., Fordyce, R.E. 2015. Enamel ultrastructure in fossil cetaceans (Cetacea: Archaeoceti and Odontoceti). *PLOS ONE*, doi: 10.1371/journal.pone.0116557.

Madsen, P.T., Lammers, M., Wisniewska, D., Beedholm, K. 2013. Nasal sound production in echolocating delphinids (*Tursiops truncatus* and *Pseudorca crassidens*) is dynamic, but unilateral: clicking on the right side and whistling on the left side. *Journal of Experimental Biology* 216: 4091-4102.

Marx, F.G., Fordyce, R.E. 2015. Baleen boom or bust: a synthesis of mysticete phylogeny, diversity and disparity. *Royal Society Open Science* 2: 140434.

———. 2016. A link no longer missing: new evidence for the cetotheriid affinities of *Caperea*. *PLOS ONE* 11(10): e10164059.

Marx, F.G., Kohno, N. 2016. A new Miocene baleen whale from the Peruvian desert. *Royal Society Open Science*, dx.doi.org: 10.6084/m9.

Marx, F.G., Tsai, C.-H., Fordyce, R.E. 2015. A new early Oligocene toothed "baleen" whale (Mysticeti: Aetiocetidae) from western North America: one of the smallest. *Royal Society Open Science* 2. doi: 10.1098/rsos.150476.

Marx, F.G., Bosselears, M.E.J., Louwye, S. 2016a. A new species of *Metopocetus* (Cetacea, Mysticeti, Cetotheriidae) from the late Miocene of the Netherlands. *PeerJ* 4: e1572. doi: 10.7717/peerj.1572.

Marx, F.G., Hocking, D.P., Park, T., Ziegler, T., Evans, A.R., Fitzgerald, E.M.G. 2016b. Suction feeding preceded filtering in baleen whale evolution. *Memoirs of the Museum Victoria* 75: 71-82.

McGowen, M.R., Gatesy, J., Wildman, D. 2015. Molecular evolution tracks macroevolutionary transitions in Cetacea. *Trends in Ecology and Evolution* 29(6): 336-346.

Mead, J., Fordyce, R.E. 2009. *The Therian Skull: A Lexicon with Emphasis on Odontocetes*. Smithsonian Contributions to Zoology, no. 627. Smithsonian Institution Scholarly Press, Washington, DC.

Mirceta, S., Signore, A.V., Burns, J.M., Cossins, A.R., Campbell, K.L., Berenbrink, M. 2013. Evolution of mammalian diving capacity traced by myoglobin net surface charge. *Science* 340: 1303.

Montgomery, S.H., Geisler, J.H., McGowen, M.R., Fox, C., Marino, L., Gatesy, J. 2013. The evolutionary history of cetacean brain and body size. *Evolution* 67(11): 3339-3353.

Muizon, C. de. 1984. *Les vertébrés fossiles de la formation Pisco (Pérou) II: les odontocetes (Cetacea, Mammalia) du Pliocene inférieur de Sud-Sacaco*, Institut Français d'Études Andines Éditions Recherche sur les Civilizations Mémoire, no. 50. Paris.

———. 1988. Les relations phylogénétiques des Delphinida (Cetacea, Mammalia). *Annales de Paleontologie* 74: 159-227.

Muizon, C. de, Domning, D.P., Ketten, D.R. 2001. *Odobenocetops peruvianus*, the walrus convergent delphinoid (Cetacea, Mammalia) from the lower Pliocene of Peru. *Smithsonian Contributions in Paleobiology* 93: 223-261.

Murakami, M., Shimada, C., Hikida, Y., Soeda, Y., Hirano, H. 2014. *Eodelphis kabatensis*, a new name for the oldest true dolphin *Stenella kabatensis* Horikawa, 1977 (Cetacea, Odontoceti, Delphinidae), from the upper Miocene of Japan, and the phylogeny and paleobiogeography of Delphinoidea. *Journal of Vertebrate Paleontology* 34(3): 491-511.

Panagiotopoulou, O., Spyridis, P., Mehari Abrahu, H., Carrier, D.R., Pataky, T.C. 2016. Architecture of the sperm whale forehead facilitates ramming combat. *PeerJ* 4: e1895. https://doi.org/10.7717/peerj.1895.

Peredo, C.M., Uhen, M.D. 2016. A new basal Chaeomysticete (Mammalia: Cetacea) from the late Oligocene Pysht Formation of Washington, U.S.A. *Papers in Paleontology* 1-22. doi: 10.1002/spp2.1051.

Post, K., Kompanje, E.J.O. 2010. A new dolphin (Cetacea, Delphinidae) from the Plio-Pleistocene of the North Sea. *Deinsea* 14: 1-13.

Pyenson, N.D., Sponberg, S.N. 2011. Reconstructing body size in extinct crown Cetacea (Neoceti) using allometry, phylogenetic methods and tests from the fossil record. *Journal of Mammalian Evolution* 18: 269-288.

Pyenson, N.D., Irmis, R.B., Lipps, J.H., Barnes, L.G., Mitchell, E.D., McLeod, S.A. 2009. Origin of a widespread marine bonebed deposited during the Miocene Climatic Optimum. *Geology* 37(6): 519-522.

Pyenson, N.D., Goldbogen, J.A., Vogl, A.W., Szathmay, G., Drake, R.L., Shadwick, R.E. 2012. Discovery of a sensory organ that coordinates lunge feeding in rorquals. *Nature* 485: 498-501.

Pyenson, N.D., Vélez-Juarbe, J., Gutstein, C.S., Little, H., Vigil, D., O'Dea, A. 2015. *Isthminia panamensis*, a new fossil inioid (Mammalia, Cetacea) from the Chagres Formation of Panama and the evolution of "river dolphins" in the Americas. *PeerJ* 3: e1227. doi: 10.7717/peerj. 1227.

Racicot, R.A., Deméré, T.A., Beatty, B.L., Boessenecker, R.W. 2014. Unique feeding morphology in a new prognathous extinct porpoise from the Pliocene of California. *Current Biology* 24: 774-779.

Racicot, R.A., Gearty, W., Kohno, N., Flynn, J.J. 2016. Comparative anatomy of the bony labyrinth of extant and extinct porpoises (Cetacea: Phocoenidae). *Biological Journal of the Linnean Society*, doi: 10.1111/bij.12857.

Sanders, A.E., Geisler, J.H. 2015. A new basal odontocete from the upper Rupelian of South Carolina, U.S.A., with contributions to the systematics of *Xenorophus* and *Microcetus* (Mammalia, Cetacea). *Journal of Vertebrate Paleontology* 35(1): e890107.

Scilla, A. 1670. *La vana speculazione disingannata dal senso: lettera risponsiva circa i Corpi Marini, che Petrificati si trovano in varij luoghi terrestri*. Andrea Colicchia, Naples, Italy.

Tanaka, Y., Fordyce, R.E. 2014. Fossil dolphin *Otekaikea marplesi* (latest Oligocene, New Zealand) expands the morphological and taxonomic diversity of Oligocene cetaceans. *PLOS ONE* 9: e107972. doi: 10.1371/journal.pone.0107972.

———. 2015a. A new Oligo-Miocene dolphin from New Zealand: *Otekaikea huata* expands diversity of early Platanistoidea. *Palaeontologica Electronica* 18: 1-71.

———. 2015b. Historically significant late Oligocene dolphin *Microcetus hectori* Benham 1935: a new species of *Waipatia* (Platanistoidea). *Journal of the Royal Society of New Zealand* 45(3): 135–150.

———. 2016. *Awamokoa tokarahi*, a new basal dolphin in the Platanistoidea (late Oligocene, New Zealand). *Journal of Systematic Palaeontology*, doi: 10.1080/1477201999 .2016.1202339.

Thewissen, J.G.M., Cohn, M.J., Stevens, L.S., Bajpai, S., Heyning, J., Horton, W.E. Jr. 2006. Developmental basis for hind-limb loss in dolphins and origin of the cetacean bodyplan. *Proceedings of the National Academy of Sciences USA* 103: 8414–8418.

Thewissen, J.G.M., Cooper, L.N., Behringrt, R.R. 2012. Developmental biology enriches paleontology. *Journal of Vertebrate Paleontology* 32: 1223–1234.

Tsai, C.-H., Ando, T. 2015. Niche partitioning in Oligocene toothed mysticetes. *Journal of Mammalian Evolution* 23(1): 33–41.

Tsai, C.-H., Fordyce, R.E. 2014. Disparate heterochronic processes in baleen whale evolution. *Evolutionary Biology* 41: 299–307.

———. 2015. The earliest gulp-feeding mysticetes (Cetacea: Mysticeti) from the Oligocene of New Zealand. *Journal of Mammalian Evolution* 22(4): 535–560.

———. 2016. Archaic baleen whale from the Kokoamu Greensand: earbones distinguish a new late Oligocene mysticete (Cetacea: Mysticeti) from New Zealand. *Journal of the Royal Society of New Zealand*, http://dx.doi.org /10.1080/03036758.2016.1156552.

Tsai, C.-H., Kohno, N. 2016. Multiple origins of gigantism in stem baleen whales. *Science in Nature* 103: 89.

Uhen, M.D. 2004. *Form, Function, and Anatomy of Dorudon atrox (Mammalia, Cetacea): An Archaeocete from the Middle to Late Eocene of Egypt*, University of Michigan, Papers on Paleontology, no. 34. University of Michigan, Ann Arbor, MI.

Vélez-Juarbe, J., Wood, A.R., De Gracia, C., Hendy, A.J.W. 2015. Evolutionary patterns among living and fossil kogiid sperm whales: evidence from the Neogene of Central America. *PLOS ONE* 10(4): e0123909. doi: 10.1371/journal. pone.0123909.

Yamato, M., Pyenson, N.D. 2015. Early development and orientation of the acoustic funnel into the evolution of sound reception pathways in cetaceans. *PLOS ONE* 10(3): e 0118582.

Chapter 4. Aquatic Carnivorans

Adam, P.J., Berta, A. 2002. Evolution of prey capture strategies and diet in the Pinnipedimorpha (Mammalia, Carnivora). *Oryctos* 4: 83–107.

Amson, E., Muizon, C. de. 2014. A new durophagous phocid (Mammalia: Carnivora) from the late Neogene of Peru and considerations on monachine seals phylogeny. *Systematic Paleontology* 12: 523–548.

Bebej, R.M. 2009. Swimming mode inferred from skeletal proportions in the fossil pinnipeds *Enaliarctos* and *Allodesmus* (Mammalia, Carnivora). *Journal of Mammalian Evolution* 16: 77–97.

Berta, A., Deméré, T.A. 1986. *Callorhinus gilmorei* n. sp. (Carnivora: Otariidae) from the San Diego Formation (Blancan) and its implications for otariid phylogeny. *Transactions of the San Diego Society of Natural History* 21: 111–126.

Berta, A., Ray, C.E. 1990. Skeletal morphology and locomotor capabilities of the archaic pinniped *Enaliarctos mealsi*. *Journal of Vertebrate Paleontology* 10: 141–157.

Berta, A., Wyss, A.R. 1994. Pinniped phylogeny. *Proceedings of the San Diego Society of Natural History* 29: 33–56.

Berta, A., Sumich, J.L., Kovacs, K.K. 2015. *Marine Mammals: Evolutionary Biology*, 3rd ed. Elsevier, San Diego, CA.

Boessenecker, R.W. 2011. New records of the fur seal *Callorhinus* (Carnivora: Otariidae) from the Plio-Pleistocene Rio Dell formation of northern California and comments on otariid dental evolution. *Journal of Vertebrate Paleontology* 31: 454–467.

Boessenecker, R.W., Churchill, M. 2013. A reevaluation of the morphology, paleoecology, and phylogenetic relationships of the enigmatic walrus *Pelagiarctos*. *PLOS ONE* 8: e54311.

———. 2015. The oldest known fur seal. *Biology Letters*, no. 2: 20140835.

———. 2016. The origin of elephant seals: implications of a fragmentary late Pliocene seal (Phocidae: Miroungini) from New Zealand. *New Zealand Journal of Geology and Geophysics*, doi: 10.1080/00288306.2016.1199437.

Brunner, S. 2004. Fur seals and sea lions (Otariidae): identification of species and taxonomic review. *Systematics and Biodiversity* 1: 339–439.

Churchill, M., Boessenecker, R.W. 2016. Taxonomy and biogeography of the Pleistocene New Zealand sea lion *Neophoca palatina* (Carnivora: Otariidae). *Journal of Paleontology*, doi: 10.1017/jpa2016.15.

Churchill, M., Clementz, M.T. 2015. Functional implications of variation in tooth spacing and crown size in Pinnipedimorpha (Mammalia: Carnivora). *Anatomical Record* 298: 878-902.

Churchill, M., Boessenecker, R.W., Clementz, M.T. 2014. The late Miocene colonization of the southern hemisphere by fur seals and sea lions (Carnivora: Otariidae). *Zoological Journal of the Linnean Society* 172(1): 200-225.

Churchill, M., Clementz, M.T., Kohno, N. 2015. Cope's rule and the evolution of body size in Pinnipedimorpha (Mammalia: Carnivora). *Evolution* 69: 201-205.

Cozzuol, M. 2001. A "northern" seal from the Miocene of Argentina: implications for phocid phylogeny and biogeography. *Journal of Vertebrate Paleontology* 21(3): 415-421.

Cullen, T.M., Fraser, D., Rybczynski, N., Schroder-Adams, C. 2014. Early evolution of sexual dimorphism and polygyny in Pinnipedia. *Evolution* 68(5): 1464-1484.

Debey, L.B., Pyenson, N.D. 2013. Osteological correlates and phylogenetic analysis of deep diving in living and extinct pinnipeds: what good are big eyes? *Marine Mammal Science* 29: 48-83.

Deméré, T.A. 1994. Two new species of fossil walruses (Pinnipedia: Odobenidae) from the upper Pliocene San Diego Formation, California. *Proceedings of the San Diego Society of Natural History* 29: 77-98.

Deméré, T.A., Berta, A. 2002. The pinniped Miocene *Desmatophoca oregonensis* Condon, 1906 (Mammalia: Carnivora) from the Astoria Formation, Oregon. *Smithsonian Contributions to Paleobiology* 93: 113-147.

———. 2005. New skeletal material of *Thalassoleon* (Otariidae: Pinnipedia) from the late Miocene-early Pliocene (Hemphillian) of California. *Florida Museum of Natural History Bulletin* 45: 379-411.

Deméré, T.A., Berta, A., Adam, P.J. 2003. Pinnipedimorph evolutionary biogeography. *Bulletin of the American Museum of Natural History* 279: 32-76.

Dewaele, L., Lambert, O., Louwye, S. 2017. On *Prophoca* and *Leptophoca* (Pinnipedia, Phocidae) from the Miocene of the North Atlantic realm: redescription, phylogenetic affinities and paleobiogeographic implications. *Peerj* 5:e3024; doi 10.7717/peerj.3024.

Fordyce, R.E. 2009. Cetacean evolution. In: Perrin, W.F., Wursig, B., Thewissen, J.G.M. (eds), *Encyclopedia of Marine Mammals*, 2nd ed. Elsevier, San Diego, CA, pp. 201-207.

Fulton, T.L., Strobeck, C. 2010. Multiple markers and multiple individuals refine true seal phylogeny and bring molecules and morphology back in line. *Proceedings of the Royal Society B* 277: 1065-1070.

Govender, R. 2015. Preliminary phylogenetic and biogeographic history of the Pliocene seal, *Homiphoca capensis* from Langebaanweg, South Africa. *Transactions of the Royal Society of South Africa* 70(1): 25-39.

Higdon, J.W., Binida-Emonds, O.R.P., Beck, R.M.D., Ferguson, S.H. 2007. Phylogeny and divergence of the pinnipeds (Carnivora: Mammalia) assessed using a multigene dataset. *BMC Evolutionary Biology* 7: 216.

Hocking, D.P., Marx, F.G., Park, T., Fitzgerald, E.M.G., Evans, A.R. 2017. A behavioural framework for the evolution of feeding in predatory aquatic mammals. *Proceedings of the Royal Society B* 284: 20162750 http://dx.doi.org/10.1098/rspb.2016.2750

Jones, K.E., Goswami, A. 2010. Quantitative analysis of the influences of phylogeny and ecology on phocid and otariid pinniped (Mammalia; Carnivora) cranial morphology. *Journal of Zoology* 280: 297-308.

Jones, K.E., Smaers, J.B., Goswami, A. 2015. Impact of the terrestrial-aquatic transition on disparity and rates of evolution in the carnivoran skull. *BMC Evolutionary Biology* 15: 8.

Kienle, S.S., Berta, A. 2015. The better to eat you with: the comparative feeding morphology of phocid seals. *Journal of Anatomy* 228(3): 396-413.

Kohno, N., Ray, C.E. 2008. Pliocene walruses from the Yorktown Formation of Virginia and North Carolina, and a systematic revision of the North Atlantic Pliocene walruses. *Virginia Museum of Natural History Special Publication* 14: 39-80.

Koretsky, I.A., Rahmat, S. 2013. First record of fossil Cystophorinae (Carnivora, Phocidae): middle Miocene seals from the northern Parathethys. *Rivista italiana di paleontolgia e stratigrafia* 119(3): 325-350.

Koretsky, I.A., Ray, C.E. 2008. Phocidae of the Pliocene of eastern USA. In: Ray, C.E., Bohaska, D., Koretsky, I.A., Ward, L.W., Barnes, L.G. (eds), *Geology and Paleontology of the Lee Creek Mine, North Carolina*, IV, vol. 15. Virginia Museum of Natural History, Special Publication. Virginia Museum of Natural History, Martinsville, VA, pp. 81-140.

Koretsky, I.A, Rahmat, S. 2015. A new species of the subfamily Devinophocinae (Carnivora, Phocidae) from the central Parathethys. *Revista Italiana di Paleontologia e Stratigrafia* 121(1): 31-47.

Koretsky, I.A., Ray, C.E., Peters, N. 2012. A new species of *Leptophoca* (Carnivora, Phocidae, Phocinae) from both sides of the North Atlantic Ocean (Miocene seals of the Netherlands, part I). *Denisea-Annual of the Natural History Museum Rotterdam* 15: 1-2.

Koretsky, I.A., Sanders, A.E. 2002. Paleontology from the late Oligocene Ashley and Chandler Bridge Formations of

South Carolina, 1: Paleogene pinniped remains: the oldest known seal (Carnivora: Phocidae). *Smithsonian Contributions in Paleobiology* 93: 179-183.

Liwanag, H.E.M., Berta, A., Costa, D.P., Abney, M., Williams, T.M. 2012a. Morphological and thermal properties of mammalian insulation: the evolution of fur for aquatic living. *Biological Journal of the Linnean Society* 106: 926-939.

Liwanag, H.E.M., Berta, A., Costa, D.P., Budge, S.M., Williams, T.M. 2012b. Morphological and thermal properties of mammalian insulation: the evolutionary transition to blubber in pinnipeds. *Biological Journal of the Linnean Society* 107: 774-787.

Loch, C., Boessenecker, R.W, Churchill, M., Kieser, J. 2016. Enamel ultrastructure of fossil and modern pinnipeds: evaluating hypotheses of feeding adaptations in the extinct walrus *Pelagiarctos*. *Nature Communications* 103: 44.

Mitchell, E.D., Tedford, R.H. 1973. The Enaliarctinae: a new group of extinct aquatic Carnivora and a consideration of the origin of the Otariidae. *Bulletin of the American Museum of Natural History* 151: 201-284.

Rybczynski, N., Dawson, M.R., Tedford, R.H. 2009. A semiaquatic Arctic mammalian carnivore from the Miocene epoch and origin of Pinnipedia. *Nature* 458: 1021-1024.

Scheel, D.-M., Slater, G.J., Kolokotronis, S.-O., Potter, C.W., Rotstein, D.S., Tsangaras, K., Greenwood, A.D., Helgen, K.M. 2014. Biogeography and taxonomy of extinct and endangered monk seals illuminated by ancient DNA and skull morphology. *ZooKeys* 409: 1-33.

Stirton, R.A. 1960. A marine carnivore from the Clallam Miocene Formation, Washington: its correlation with nonmarine faunas. *University of California Publications in the Geological Sciences* 36(7).

Tanaka, Y., Kohno, N. 2015. A new late Miocene odobenid (Mammalia: Carnivora) from Hokkaido, Japan suggests rapid diversification of basal Miocene odobenids. *PLOS ONE*, e0131856. doi: 10.1371/journal.pone.0131856.

Tedford, R.H., Barnes, L.G., Ray, C.E. 1994. The early Miocene littoral ursoid carnivoran *Kolponomos*: systematics and mode of life. *Proceedings of the San Diego Society of Natural History* 29: 11-32.

Tseng, Z.J., Grohé, C., Flynn, J.J. 2016. A unique feeding strategy of the extinct marine mammal *Kolponomos*: convergence on sabretooths and sea otters. *Proceedings of the Royal Society B* 283(1826). doi: 10.1098/rspb.2016.0044.

Valenzuela-Toro, A.M., Pyenson, N.D., Gutstein, C.S., Suarez, M.E. 2016. A new dwarf seal from the late Neogene of South America and the evolution of pinnipeds in the southern hemisphere. *Papers in Paleontology* 2(1): 101-115.

Velez-Juarbe, J. 2017. *Eotaria citrica*, sp. nov., a new stem otariid from the "Topanga" Formation of southern California. *Peerj* 5:e3022 https://doi.org/10.7717/peerj.3022.

Wallace, D.R. 2007. *Neptune's Ark: From Ichthyosaurs to Orcas*. University of California Press, Berkeley, CA.

Wyss, A.R. 1994. The evolution of body size in phocids: some ontogenetic and phylogenetic observations. *Proceedings of the San Diego Society of Natural History* 29: 69-75.

Chapter 5. Crown Sirenians and Their Desmostylian Relatives

Balaguer, J., Alba, D.M. 2016. A new dugong species (Sirenia, Dugongidae) from the Eocene of Catalonia (NE Iberian Peninsula). *Comptes Rendus Palevol* 13: 489-500.

Barnes, L.G. 2013. A new genus and species of late Miocene paleoparadoxiid (Mammalia, Desmostylia) from California. *Natural History Museum of Los Angeles County Contributions in Science* 521: 51-114.

Beatty, B.L. 2009. New material of *Cornwallius sookensis* (Mammalia: Desmostylia) from the Yaquina Formation of Oregon. *Journal of Vertebrate Paleontology* 29: 894-909.

Beatty, B.L., Cockburn, T.C. 2015. New insights on the most primitive desmostylians from a partial skeleton of *Behemotops* (Desmostylia, Mammalia) from Vancouver Island, British Columbia. *Journal of Vertebrate Paleontology*, doi: 10.1080/02724634.2015.979939.

Clementz, M.T., Hoppe, K.A., Koch, P.L. 2003. A paleoecological paradox: the habitat and dietary preferences of the extinct tethythere *Desmostylus*, inferred from stable isotope analysis. *Paleobiology* 29(4): 506-519.

Clementz, M.T., Sorbi, S., Domning, D.P. 2009. Evidence of Cenozoic environmental and ecological change from stable isotope analysis of sirenian remains from the Tethys-Mediterranean region. *Geology* 37: 307-310.

Crerar, L.L., Crerar, A.P., Domning, D.P., Parsons, E.C.M. 2014. Rewriting the history of an extinction: was a population of Steller's sea cows (*Hydrodamalis gigas*) at St. Lawrence Island also driven to extinction? *Biology Letters* 10. doi: 10.1098/rsbl.2014.0878.

Domning, D.P. 2002. The terrestrial posture of desmostylians. *Smithsonian Contributions in Paleobiology* 93: 99-111.

———. 2005. Sirenia of the West Atlantic and Caribbean region, VII: Pleistocene *Trichechus manatus* Linnaeus, 1758. *Journal of Vertebrate Paleontology* 25(3): 685-701.

Fitzgerald, E.M.G., Vélez-Juarbe, J., Wells, R.T. 2013. Miocene sea cow (Sirenia) from Papua New Guinea sheds light on sirenian evolution in the Indo-Pacific. *Journal of Vertebrate Paleontology* 33(4): 956-963.

Fordyce, R.E. 2009. Fossil sites, noted. In: Perrin, W.F., Wursig, B., Thewissen, J.G.M. (eds), *Encyclopedia of Marine Mammals*, 2nd ed. Elsevier, San Diego, CA, pp. 459-466.

Gheerbrant, E., Domning, D.P., Tassy, P. 2005. Paenungulata (Sirenian, Proboscidea, Hyracoidea, and relatives. In: Rose, K.D., Archibald, J.D. (eds), *The Rise of Placental Mammals: Origins and Relationships of the Major Extant Clades*. Johns Hopkins University Press, Baltimore, MD, pp. 84-105.

Hayashi, S., Houssaye, A., Nakajima, Y., Chiba, K., Ando, T., Sawamura, H., Inunzuka, N., Kaneko, N., Osaki, T. 2013. Bone inner structure suggests increasing aquatic adaptations in Desmostylia (Mammalia, Afrotheria). *PLOS ONE* 8: e59146.

Inuzuka, N. 1984. Skeletal restoration of the desmostylians: herpetiform mammals. *Memoirs of the Faculty of Science Kyoto University Series Biology* 9: 157-253.

Marsh, H., Beck, C.A., Vargo, T. 1999. Comparison of the capabilities of dugongs and West Indian manatees to masticate seagrasses. *Marine Mammal Science* 15: 250-255.

Reinhart, R. 1959. A review of the Sirenia and Desmostylia. *University of California Publications in the Geological Sciences* 36(1): 1-145.

Rommel, S., Reynolds, J.E. III. 2009. Skeleton, postcranial. In: Perrin, W.F., Wursig, B., Thewissen, J.G.M. (eds), *Encyclopedia of Marine Mammals*, 2nd ed. Elsevier, San Diego, CA, pp. 1021-1033.

Vélez-Juarbe, J., Domning, D.P. 2015. Fossil Sirenia of the West Atlantic and Caribbean region, XI: *Callistosiren boriquensis*, gen. et sp. nov. *Journal of Vertebrate Paleontology* 35(1): e885034. doi: 10.1080/02724634.2014.885034.

Vélez-Juarbe, J., Domning, D.P., Pyenson, N.D. 2012. Iterative evolution of sympatric seacow (Dugongidae, Sirenia) assemblages during the past ~26 million years. *PLOS ONE* 7: e31294.

Zalmout, I.S., Gingerich, P.D. 2012. Late Eocene sea cows (Mammalia, Sirenia) from Wadi Al Hitan in the western Desert of Fayum, Egypt. *Papers of the Michigan Museum of Paleontology*, no. 37.

Chapter 6. Aquatic Sloths and Recent Occupants of the Sea, Sea Otters and Polar Bears

Amson, E., Argot, C., McDonald, H.G., Muizon, C. de. 2014. Osteology and functional morphology of the hind limb of the marine sloth *Thalassocnus* (Mammalia, Tardigrada). *Journal of Mammalian Evolution* 22(3): 355-419.

———. 2015a. Osteology and functional morphology of the axial postcranium of the marine sloth *Thalassocnus* (Mammalia, Tardigrada) with paleobiological implications. *Journal of Mammalian Evolution* 22(4): 473-518.

———. 2015b. Osteology and functional morphology of the forelimb of the marine sloth *Thalassocnus* (Mammalia, Tardigrada). *Journal of Mammalian Evolution* 22(2): 169-241.

Amson, E., Muizon, C. de, Gaudin, T.J. 2016. A reappraisal of the phylogeny of the Megatheria (Mammalia: Tardigrada), with emphasis on the relationships of the Thalassocninae, the marine sloths. *Zoological Journal of the Linnean Society*, doi: 10.1111/zoj.12450.

Berta, A., Morgan, G.S. 1985. A new sea otter (Carnivora: Mustelidae) from the late Miocene and early Pliocene (Hemphillian) of North America. *Journal of Paleontology* 59: 809-819.

Cahill, J.A., Green, R.E., Fulton, T.L., Stiller, M., Jay, F., Ovasyanikov, N., Salamzade, R., St John, J., Stirling, I., Slatkin, M., Shapiro, B. 2013. Genomic evidence for island population conversion resolves conflicting theories of polar bear evolution. *PLOS Genetics* 9: e10003345.

Estes, J.A., Bodkin, J.L. 2002. Otters. In: Perrin, W.F., Wursig, B., Thewissen, J.G.M. (eds), *Encyclopedia of Marine Mammals*, 1st ed. Academic Press, San Diego, CA, pp. 842-855.

Fordyce, R.E. 2009. Fossil sites, noted. In: Perrin, W.F., Wursig, B., Thewissen, J.G.M. (eds), *Encyclopedia of Marine Mammals*, 2nd ed. Elsevier, San Diego, CA, pp. 459-466.

Geraads, D., Alemseged, Z., Bobe, R., Reed, D. 2011. *Enhydriodon dikikae*, sp. nov. (Carnivora: Mammalia), a gigantic otter from the Pliocene of Dikika, Lower Awash, Ethiopia. *Journal of Vertebrate Paleontology* 31: 447-453.

Ingolfsson, O., Wiig, Ø. 2008. Late Pleistocene find in Svalbard: the oldest remains of a polar bear (*Ursus maritimus* Phipps, 1744) ever discovered. *Polar Research* 28: 455-462.

Lambert, W.D. 1997. The osteology and paleoecology of the giant otter *Enhydritherium terraenovae*. *Journal of Vertebrate Paleontology* 17: 738-749.

Lindqvist, C., Schuster, S.C., Sun, Y., Talbot, S.L., Qi, J., Ratan, A., Tomsco, L.P., Kasson, L., Zeyl, E., Aars, J., Miller, W., Ingólfsson, Ó., Bachmann, L., Wiig, Ø. 2010. Complete mitochondrial genome of a Pleistocene jawbone unveils the origin of polar bear. *Proceedings of the National Academy of Sciences USA* 107: 5053-5057.

Liu, S., Lorenzen, E.D., Fumagalli, M., Li, B., Harris, K., Xiong, Z., Zhou, L., Mead, J.I., Spies, A.E., Sobolik, K.D. 2000. Skeleton of extinct North American sea mink. *Quaternary Research* 53: 247-262.

Liu, S., Lorenzen, E.D., Fumagalli, M., Li, B., Harris, K., Xiong, Z., Zhou, L., Korneliussen, T.S., Somel, M., Babbitt, C., Wray, G., Li, J., He, W., Wang, Z., Fu, W.,

Xiang, X., Morgan, C.C., Doherty, A., O'Connell, M. J., McInerney, J.O., Born, E.W., Dalén, L., Dietz, R., Orlando, L., Sonne, C., Zhang, G., Nielsen, R. Willerslev, Wang, J. 2014. Population genomics reveal recent speciation and rapid evolutionary adaptation in polar bears. *Cell* 157: 785-794.

Mead, J.I., Spies, A.E., Sobolik, K.D. 2000. Skeleton of extinct North American sea mink. *Quaternary Research* 53: 247-262.

Muizon, C. de, McDonald, H.G. 1995. An aquatic sloth from the Pliocene of Peru. *Nature* 375: 224-227.

Muizon, C. de, McDonald, H.G., Salas, R., Urbina, M. 2003. A new early species of the aquatic sloth *Thalassocnus* (Mammalia, Xenarthra) from the late Miocene of Peru. *Journal of Vertebrate Paleontology* 23: 886-894.

———. 2004a. The evolution of feeding adaptations of the aquatic sloth *Thalassocnus*. *Journal of Vertebrate Paleontology* 24: 398-410.

———. 2004b. The youngest species of the aquatic sloth *Thalassocnus* and a reassessment of the relationships of the nothrothere sloths (Mammalia: Xenarthra). *Journal of Vertebrate Paleontology* 24: 287-397.

Pickford, M. 2007. Revision of the Mio-Pliocene bunodont otter-like mammals of the Indian subcontinent. *Estudios Geológicos* 63(1): 83-127.

Timm-Davis, L.L., DeWitt, T.J., Marshall, C.D. 2015. Divergent skull morphology supports two trophic specializations in Otters (Lutrinae). *PLOS ONE* 9(10): e0143236.

Willemsen, G.F. 1992. A revision of the Pliocene and Quaternary Lutrinae from Europe. *Scripta Geologica* 101: 1-115.

Chapter 7. Diversity Changes through Time

Ainley, D., Ballard, G., Ackley, S., Blight, L.K., Eastman, J.T., Emslie, S.D., Lescroel, A., Olmanstron, S., Townsend, S.E., Tynan, C.T., Wilson, P., Woehler, E. 2007. Paradigm lost, or is top-down forcing no longer significant in the Antarctic marine ecosystem? *Antarctic Science* 19: 283-290.

Alter, S.E., Meyer, M., Post, K., Czechowski, P., Gravlund, P., Gaines, C., Rosenbaum, H., Kaschner, K., Turvey, S.T., Van Der Plicht, J., Shapiro, B., Hofreiter, M. 2015. Climate impacts on transocean dispersal and habitat in gray whales from the Pleistocene to 2100. *Molecular Ecology* 24: 1510-1522.

Ballance, L.T., Pitman, R.L., Hewitt, R.P., Siniff, D.B., Trivleplece, W.Z., Clapham, P.J., Brownell, R.L. Jr. 2006. The removal of large whales from the Southern Ocean: evidence for long term ecosystem effects? In: Estes, J.A., Demaster, D.P., Doak, D.F., Williams, T.M., Brownell, R.L.

Jr. (eds), *Whales, Whaling, and Ocean Ecosystems.* University of California Press, Berkeley, CA, pp. 215-230.

Benton, M.J. 2009. The Red Queen and Court Jester: species diversity and the role of biotic and abiotic factors through time. *Science* 323: 728-732.

Braje, T.H., Rick, T.C. (eds). 2011. *Human Impacts on Seals, Sea Lions, and Sea Otters.* University of California Press, Berkeley, CA.

Collins, C.J., Rawlence, N.J., Proust, S., Anderson, C.N.K., Knapp, M., Scofiled, R.P., Robertson, B.C., Smith, I., Matisoo-Smith, E.A., Chilvers, B.L., Waters, J.M. 2014. Extinction and recolonization of coastal megafauna following human arrival in New Zealand. *Proceedings of the Royal Society B*, doi: 10.1098/rspb.2014.0097.

Danise, S., Domenici, S. 2014. A record of shallow-water whale falls from Italy. *Lethaia* 47: 229-243.

Davidson, A.D., Boyer, A.G., Kim, H., Pompa-Mansilla, S., Hamilton, M.J., Costa, D.P., Ceballos, G., Brown, J.H. 2012. Drivers and hotspots of extinction risk in marine mammals. *Proceedings of the National Academy of Sciences USA* 109: 3395-3400.

De Bruyn, M., Hoelzel, A.R., Carvalho, G.R., Hofreiter, M. 2011. Faunal histories from Holocene ancient DNA. *Trends in Ecology and Evolution* 26(8): 405-413.

De Master, D.P., Trites, A.W., Clapham, P., Mizroch, S., Wade, P., Small, R.J., Ver Hoef, J. 2006. The sequential megafaunal collapse hypothesis: testing with existing data. *Progress in Oceanography* 68: 329-342.

Estes, J.A., Tinker, M.T., Williams, T.M., Doak, D.F. 1998. Killer whale predation on sea otters linking oceanic and nearshore ecosystems. *Science* 282: 473-476.

Estes, J.A., Terborgh, J., Brashares, J.S., Power, M.E., Berger, J., Bond, W.J., Carpenter, S.R., Essington, T.E., Holt, R.D., Jackson, J.B.C., Marquis, R.J., Oksanen, L., Okansen, T., Paine, R.T., Pikitch, E.K., Ripple, W.J., Sandin, S.A., Soule, M.E., Virtanen, R., Wardle, D.A. 2011. Trophic downgrading of planet Earth. *Science* 333: 301-306.

Estes, J.A., Burdin, A., Doak, D.F. 2015. Sea otters, kelp forests, and the extinction of Steller's sea cow. *Proceedings of the National Academy of Sciences USA* 113(4): 880-885.

Finnega, S., Anderson, S.C., Harnik, P.G., Simpson, C., Tittensour, D.P., Brynes, J.F., Finkel, Z.F., Lindberg, D.R., Liow, L.H., Lockwood, H.K., McClain, C.R., McGuire, J.L., O'Dea, A., Pandolfi, J.M. 2015. Paleontological baselines for evaluating extinction risk in the modern oceans. *Science* 348: 567-570.

Foote, A.D., Kaschner, K., Schultze, S.E., Garilao, C., Ho, S.Y.W., Klaas, P., Higham, T.F.G., Stokowska, C.B., van der Es, H., Embling, C.B., Gregersen, K., Johansson, F., Willerslev, E., Gilbert, M.T.P. 2013. Ancient DNA reveals that bowhead whale lineages survived Late Pleistocene climate change and habitat shifts. *Nature Communications* 4: 1677. doi: 10.1038/ncommuns2714.

Harnik, P.G., Lotze, H.K., Anderson, S.C., Finkel, Z.V., Finnegan, S., Lindberg, D.R., Liow, L.H., Lockwood, R., McClain, C.R., McGuire, J.L., O'Dea, A., Pandolfi, J.M., Simpson, C., Tittensor, D.P. 2012. Extinctions in ancient and modern seas. *Trends in Ecology and Evolution* 27(11): 608-617.

Hunt, K.E., Stimmelayr, R., George, C., Suydam, R., Brown, H. Jr., Rolland, R.M. 2014. Baleen hormones: a novel assessment of stress and reproduction in bowhead whales. *Conservation Physiology* 2. doi: 10.1093/conphys/cou030.

Jackson, J.B.C., Kirby, M.X., Berger, W.H., Bjorndal, K.A., Botsford, L.W., Bourque, B.J., Bradbury, R.H., Cooke, R., Erlandson, J., Estes, J.A., Hughes, T.P., Kidwell, S., Lange, C.B., Lenihan, H.S., Pandolfi, J.M., Peterson, C.H., Steneck, R.S., Tegner, M.J., Warner, R.R. 2001. Historical overfishing and the recent collapse of coastal ecosystems. *Science* 293: 629-638.

Kaschner, K., Tittensor, D.P., Ready, J., Gerrodette, T., Worm, B. 2011. Current and future patterns of global marine mammal biodiversity. *PLOS ONE*, e19653. doi: 10.1371/journal.pone.0019653.

Koch, P.L., Fox-Dobbs, K., Newsome, S.D. 2009. The isotopic ecology of fossil vertebrates and conservation paleobiology. *Paleontological Society Papers* 15: 95-112.

Lindberg, D.R., Pyenson, N.P. 2006. Evolutionary patterns in Cetacea: fishing up prey sizes through deep time. In: Estes, J.A., Demaster, D.P., Doak, D.F., Williams, T.M., Brownell, R.L. Jr. (eds), *Whales, Whaling, and Ocean Ecosystems*. University of California Press, Berkeley, CA, pp. 68-82.

Moore, S.E. 2008. Marine mammals as ecosystem sentinels. *Journal of Mammalogy* 89(3): 534-540.

Molnar, P.K., Derocher, A.E., Klanjscek, T., Lewis, M.A. 2011. Predicting climate change impacts on polar bear litter size. *Nature Communications* 2: 186.

Morin, P.A., Parsons, K.M, Archer, F.I., Ávila-Arcos, M.C., Barrett-Lennard, L.G., Dalla Rosa, L., Duchêne, S., Durban, J.W., Ellis, G.M., Ferguson, S.H., Ford, J.K., Ford, M.J., Garilao, C., Gilbert, M.T., Kaschner, K., Matkin, C.O., Petersen, S.D., Robertson, K.M., Visser, I.N., Wade, P.R., Ho, S.Y., Foote, A.D. 2015. Geographic and temporal dynamics of a global radiation and diversification in the killer whale. *Molecular Ecology* 24: 3964-3979.

Newsome, S.D., Etnier, M.A., Gifford-Gonzalez, D., Phillips, D.L., van Tuinen, M., Hadly, E.A., Costa, D.P., Kennett, D.J., Guilderson, T.P., Koch, P.L. 2007. The shifting baseline of northern fur seal ecology in the northeast Pacific Ocean. *Proceedings of the National Academy of Sciences USA* 104(23): 9709-9714.

Newsome, S.D., Clementz, M.T., Koch, P.L. 2010. Using stable isotope biogeochemistry to study marine mammal ecology. *Marine Mammal Science* 26(3): 509-572.

Pinsky, M.L., Newsome, S.D., Dickerson, B.R., Fang, Y., Van Tuinen, M., Kennett, D.J., Ream R.R., Hadly, E.A. 2010. Dispersal provided resilience to range collapse in a marine mammal: insights from the past to inform conservation biology. *Molecular Ecology* 19: 2418-2429.

Pyenson, N.P. 2011. The high fidelity of the cetacean stranding record: insights into measuring diversity by integrating taphonomy and macroecology. *Proceedings of the Royal Society B* 278: 3608-3616.

Pyenson, N.P., Lindberg, D.R. 2011. What happened to gray whales during the Pleistocene? The ecological impact of sea-level change on benthic feeding areas in the North Pacific Ocean. *PLOS ONE* 6(7): e21295.

Pyenson, N.P., Gutstein, C.S., Parham, J.F., Le Roux, J.P., Chavarria, C.C., Little, H., Metallo, A., Rossi, V., Valenzuela-Toro, A., Vélez-Jarbe, J., Santelli, C.M., Rodgers, D.R., Cozzuol, M.A., Suarez, M.E. 2014. Repeated mass strandings of Miocene marine mammals from the Atacama Region of Chile point to sudden death at sea. *Proceedings of the Royal Society B* 277: 3097-3104.

Ramp, C., Delarue, J., Palsboll, P.J., Sears, R., Hammond, P.S. 2015. Adapting to a warmer ocean-seasonal shift of baleen whale movements over three decades. *PLOS ONE*, doi: 10.1371/journal.pone.0121374.

Ruegg, K.C., Anderson, E.C., Baker, C.S., Vant, M., Jackson, J.A., Palumbi, S.R. 2010. Are Antarctic minke whales unusually abundant because of 20th century whaling. *Molecular Ecology* 19: 281-291.

Sarrazin, F., Lecomte, J. 2016. Evolution in the Anthropocene. *Science* 351(6276): 922-923.

Slater, G.J., Price, S.A., Santinia, F., Alfaro, M.J. 2010. Diversity vs disparity and the radiation of modern cetaceans. *Proceedings of the Royal Society B* 277(1697): 3097-3104.

Smith, C.R., Glover, A.G., Treude, T., Higgs, N.D., Amon, D.J. 2015. Whale-fall ecosystems: recent insights into ecology, paleoecology, and evolution. *Annual Reviews of Marine Science* 7: 571-596.

Springer, A.M., Estes, J.A., Vliet, G.B. van, Williams, T.M., Doak, D.F., Danner, E.M., Forney, K.A., Pfister, B. 2003.

Sequential megafaunal collapse in the North Pacific Ocean: an ongoing legacy of industrial whaling? *Proceedings of the National Academy of Sciences USA* 100(21): 12,223-12,228.

Springer, A.M., Estes, J.A, Vliet, G.B. van, Williams, T.M., Doak, D.F., Danner, E.M., Pfister, B. 2008. Mammal-eating killer whales, industrial whaling, and the sequential megafaunal collapse in the North Pacific Ocean: a reply to critics of Springer *et al.* 2003. *Marine Mammal Science* 24(2): 414-442.

Steeman, M.E., Hebsgaard, M.B., Fordyce, R.E., Ho, S.Y.W., Rabosky, D.L., Nielsen, R., Rahbek, C., Glenner, H., Sørensen, M.V., Willerslev, E. 2009. Radiation of extant cetaceans driven by restructuring of the ocean. *Systematic Biology* 58: 573-585.

Sydeman, W.J., Poloczanska, E., Reed, T.E., Thompson, S.A. 2015. Climate change and marine vertebrates. *Science* 356: 772-777.

Thomas, H.W., Barnes, L.G. 2015. The bone pathology osteochondrosis in extant and fossil marine mammals. *Los Angeles County Museum Contributions in Science* 523: 1-35.

Trumble, S.J., Robinson, E.M., Berman-Kowalewski, M., Potter, C.W., Usenko, S. 2013. Blue whale earplug reveals lifetime contaminant exposure and hormone profiles. *Proceedings of the National Academy of Sciences USA* 110: 16,922-16,926.

Wittemann, T.A., Izzo, C., Doubleday, Z.A., McKenzie, J., Delean, S., Gillanders, B.M. 2016. Reconstructing climate-growth relations from teeth of a marine mammal. *Marine Biology* 163: 71.

Younger, J.L., Emmerson, L.M., Miller, K.J. 2016. The influence of historical climate changes on Southern Ocean marine predator populations: a comparative analysis. *Global Change Biology* 22: 474-493.

Index